INTERNATIONAL UNION OF CRYSTALLOGRAPHY
BOOK SERIES

IUCr BOOK SERIES COMMITTEE

IUCr Monographs on Crystallography

1. *Accurate molecular structures*
   A. Domenicano, I. Hargittai, editors
2. *P. P. Ewald and his dynamical theory of X-ray diffraction*
   D.W.J. Cruickshank, H.J. Juretschke, N. Kato, editors
3. *Electron diffraction techniques, Vol. 1*
   J. M. Cowley, editor
4. *Electron diffraction techniques, Vol. 2*
   J. M. Cowley, editor
5. *The Rietveld method*
   R.A. Young, editor
6. *Introduction to crystallographic statistics*
   U. Shmueli, G.H. Weiss
7. *Crystallographic instrumentation*
   L.A. Aslanov, G.V. Fetisov, G.A.K. Howard
8. *Direct phasing in crystallography*
   C. Giacovazzo
9. *The weak hydrogen bond*
   G.R. Desiraju, T. Steiner
10. *Defect and microstructure analysis by diffraction*
    R.L. Snyder, J. Fiala and H.J. Bunge
11. *Dynamical theory of X-ray diffraction*
    A. Authier
12. *The chemical bond in inorganic chemistry*
    I.D. Brown

13 *Structure determination from powder diffraction data*
W.I.F. David, K. Shankland, L.B. McCusker, Ch. Baerlocher, editors
14 *Polymorphism in molecular crystals*
J. Bernstein
15 *Crystallography of modular materials*
G. Ferraris, E. Makovicky, S. Merlino
16 *Diffuse x-ray scattering and models of disorder*
T.R. Welberry
17 *Crystallography of the polymethylene chain: an inquiry into the structure of waxes*
D.L. Dorset
18 *Crystalline molecular complexes and compounds: structure and principles*
F.H. Herbstein

IUCr Texts on Crystallography

1 *The solid state*
A. Guinier, R. Julien
4 *X-ray charge densities and chemical bonding*
P. Coppens
5 *The basics of crystallography and diffraction, second edition*
C. Hammond
6 *Crystal structure analysis: principles and practice*
W. Clegg, editor
7 *Fundamentals of crystallography, second edition*
C. Giacovazzo, editor
8 *Crystal structure refinement: a crystallographer's guide to SHELXL*
P. Müller, editor

IUCr Crystallographic Symposia

1 *Patterson and Pattersons: Fifty years of the Patterson Junction*
J. P. Glusker, B. K. Patterson, and M. Rossi, *editors*
2 *Molecular structure: Chemical reactivity and biological activity*
J. J. Stezowski, J. Huang, and M, Shao, *editors*
3 *Crystallographic computing 4: Techniques and new technologies*
N. W. Isaacs and M. R. Taylor, *editors*
4 *Organic crystal chemistry*
J. Garbarczyk and D. W. Jones, *editors*
5 *Crystallographic computing 5: From chemistry to biology*
D. Moras, A. D. Podjarny, and J. C. Thierry, *editors*
6 *Crystallographic computing 6: A window on modern crystallography*
H. D. Flack, L. Parkanyi, and K. Simon, *editors*
7 *Correlations, transformation, and interactions in organic crystal chemistry*
D. W. Jones and A. Katrusiak, *editors*

Crystal Structure Refinement

A Crystallographer's Guide to SHELXL

Peter Müller

Massachusetts Institute of Technology, Cambridge, USA

Regine Herbst-Irmer

University of Göttingen, Germany

Anthony L. Spek

Utrecht University, The Netherlands

Thomas R. Schneider

The FIRC Institute of Molecular Oncology, Biocrystallography, and Structural Bioinformatics, Italy

Michael R. Sawaya

University of California, Los Angeles, USA

Edited by

Peter Müller

INTERNATIONAL UNION OF CRYSTALLOGRAPHY

OXFORD

UNIVERSITY PRESS

OXFORD
UNIVERSITY PRESS

Great Clarendon Street, Oxford OX2 6DP

Oxford University Press is a department of the University of Oxford.
It furthers the University's objective of excellence in research, scholarship,
and education by publishing worldwide in

Oxford New York

Auckland Cape Town Dar es Salaam Hong Kong Karachi
Kuala Lumpur Madrid Melbourne Mexico City Nairobi
New Delhi Shanghai Taipei Toronto

With offices in

Argentina Austria Brazil Chile Czech Republic France Greece
Guatemala Hungary Italy Japan Poland Portugal Singapore
South Korea Switzerland Thailand Turkey Ukraine Vietnam

Oxford is a registered trade mark of Oxford University Press
in the UK and in certain other countries

Published in the United States
by Oxford University Press Inc., New York

First published 2006

Reprinted 2007

British Library Cataloguing in Publication Data
Data available

Library of Congress Cataloging in Publication Data
Crystal structure refinement : a crystallographer's guide :
SHELXL / Peter Müller . . . [et al.]; edited by Peter Müller.
p. cm.—(International Union of Crystallography monographs on crystallography ; 19)
Includes bibliographical references.
ISBN-13: 978–0–19–857076–9 (alk. paper)
ISBN-10: 0–19–857076–7 (alk. paper)
1. Crystals—Structure. 2. Crystals—Refining. 3. Crystals—Data processing.
I. Müller, Peter, 1970– II. Series.
QD921.C772 2006
548′.81—dc22

2006011794

Typeset by Newgen Imaging Systems (P) Ltd., Chennai, India
Printed in Great Britain
on acid-free paper by
Biddles Ltd, King's Lynn, Norfolk

ISBN 978–0–19–857076–9

3 5 7 9 10 8 6 4 2

Foreword

A Short History of SHELX

The 5000 lines of FORTRAN code that became known as SHELX-76 had their origins around 1970 when the University of Cambridge replaced the ICL Titan computer with an IBM-370. My previous attempts to write programs used Titan Autocode, a simple but efficient programming language closer to assembler than to a modern high-level language. With the IBM computer came two major innovations: a FORTRAN compiler and punched cards. Being forced to rewrite my first attempt at a crystallographic least-squares refinement program (called NOSQUARES) in another language was a good opportunity to learn from my mistakes, but—since I was too lazy to read the FORTRAN manual or attend a course—I rewrote the program in a very simple subset of FORTRAN that bore a curious resemblance to Titan Autocode, and avoided features that might have been difficult to port to other computers so that I would never have to rewrite it again. This had the advantage that it produced efficient code, essential in view of the limited speed and memory of the mainframe computers of the time (about 0.0001 times that of current PCs). Actually SHELX-76 still compiles and runs correctly using almost any modern FORTRAN-95 compiler.

At the time I would have regarded myself as an inorganic chemist who was interested in applying a variety of physical methods; the title of my Ph.D. thesis (under the supervision of Evelyn Ebsworth) was 'NMR Studies of Inorganic Hydrides'. When I moved to the Georg-August University of Göttingen in 1978 I discovered that my German colleagues were so much better at 'cooking' (preparative chemistry) than I was that it would be better if I concentrated on crystal structure determination, for which there was a pressing need in order to characterize all the compounds they were synthesizing.

One of the methods we made good use of in the 1960s, for example, for determining the structures of relatively unstable -SiH_3 derivatives that had a habit of exploding in contact with air, was gas phase electron diffraction. This required synthesizing the compounds in Cambridge and taking them to Glasgow University and later to UMIST (Manchester) where Durward Cruickshank had the only operational gas phase electron diffraction machine in the country. On one visit I mentioned to Durward that I would need to do some X-ray crystallography because not all our samples were volatile enough to determine the structures in the gas phase. I had managed to find an X-ray generator and a Weissenberg camera but still needed to write a suitable Autocode program for the Titan computer to analyze the data. Durward very kindly provided me with a set of notes describing least-squares refinement that he later presented at the 1969 computing school in Ottawa: these form the basis of the least-squares refinement in SHELX to this day.

SHELX-76

SHELX was written for use by myself and my students and I never imagined that it would ever find use outside the ivory towers of Cambridge. However, after a few years during which the program was fairly well debugged, it became clear that it would be a good idea to have one definitive 'export' version. This was named SHELX-76 and was intended to be the final definitive version. SHELX-76 included Lp and absorption corrections for Weissenberg data; in addition to the camera we had acquired a Weissenberg geometry diffractometer for which I wrote the control program in binary to get the most out of the 4K 12bit words of memory. Fortunately the use of the concept of *direction cosines* made it possible to handle data from other sources. In SHELX-76 this was followed by the merging of the data to produce a list of unique reflections, some primitive direct and Patterson methods for structure solution, least-squares structure refinement, calculation of dependent parameters and Fourier syntheses. The resulting program was so large that the 5000 FORTRAN statements were too heavy to carry around as individual punched cards, so I wrote a little compression program for the FORTRAN and another one for the test data (it averaged 9 reflections per card, i.e ~9 bytes per reflection, but compromized a little on precision). The program, test data and (uncompressed) FORTRAN decompression program all fitted into a standard 2000-card box that could be posted and came with me on trips abroad. In fact these 'compressed data' can still be read using 'HKLF 1' and were subject to a brief renaissance when BITNET was introduced. Once when I was on vacation one of my students dropped the only dataset from a valuable crystal and distributed the cards all over the floor, but succeeded in cracking the code and putting the cards back in the right order before I returned!

One problem that soon became apparent was that the restriction of the array dimensions to allow only 160 atoms (including hydrogen atoms) was a little on the small side; I had thought that this would never need increasing. Dobi Rabinovitch worked out how to increase the arrays to hold 400 atoms and this became the standard version. When I was faced with the problem of converting the program to the first (Data General) minicomputers I was able to overcome the memory restrictions (the program and operating system had to fit into 64 Kbytes) by extensive use of 'overlay' (only holding a small part of the executable code in memory) and with a rather efficient blocked cascade least-squares refinement algorithm that refined the structure in small dynamically selected blocks, but only recalculated the structure factor contributions for atoms that had changed in the previous cycle. This was the basis of the XLS refinement program in the SHELXTL version that I had adapted for Syntex (who later became Nicolet, then Siemens and finally Bruker). XLS only fitted into the Nova computer memory with two bytes to spare, so further extension and even bug-fixing were difficult.

SHELX-97

I learnt a great deal about direct methods of structure solution at the excellent schools that Michael Woolfson and Lodovico Riva di Sanseverino organized, first in Parma

(1970) and then from 1974 on in Erice. By the 1980s direct methods had made such progress that I decided to produce a separate structure solution program (SHELXS-86). This was eventually followed by a new refinement program SHELXL in 1993, partly because Syd Hall and the editors of Acta Crystallographica were pestering me to produce CIF output. However CIF is by no means the ideal answer to the data exchange and archiving problem; even though the CIF file is longer than the corresponding SHELXL .res file, it lacks much information, for example about the constraints and restraints applied in the refinement. Both SHELXS and SHELXL were updated again in 1997 and proved sufficiently reliable that no further updates were required. Both programs (and the subsequent SHELXC, SHELXD and SHELXE for macromolecular phasing) were tested for many years before they were released, with the result that they were by that stage already fairly well debugged. This contrasts with the current general programming philosophy that code is distributed as soon as possible and the users will find the bugs! This package, which included the program CIFTAB for working with CIF format files and SHELXPRO that acted as an interface to the macromolecular world, became known as SHELX-97.

Documentation is always a problem, so I sent out the first beta-test copies of this package (starting in 1992) one at a time. A potential beta-tester was sent a copy of the manual and was told that he or she would be sent the programs only after sending me at least three errors in the documentation or good suggestions for improving it. I then made all the corrections before sending it to the next guinea-pig. The first testers ran it through their spelling checkers and found plenty of mistakes (my spelling was never very good) but after a couple of hundred beta-tests had been sent out, people began to complain that it was all a diabolical plot and that I had simply written an error-free manual for programs that I had no intention of sending out and that probably didn't even exist!

Program Style

Few computer programs of the antiquity of SHELX are in wide use today (though ORTEP is an even older survivor). One possibility is that the use of a very simple standard subset of FORTRAN, true even of the more recent additions to the SHELX system, makes it trivial to port the programs to new computer hardware. In comparison with other computer languages, FORTRAN has remained remarkably stable and upwards compatible. In the meantime I have learnt some C and C++ and even (several years ago) held PASCAL courses, but consider that FORTRAN is still the language of choice for rugged number-crunching programs. FORTRAN shows no signs of fading out, as exemplified by the excellent selection of FORTRAN compilers available for Linux systems, and the sheer inertia of the vast base of scientific FORTRAN code will ensure its survival for a long time to come. There are many excellent numerical libraries available for FORTRAN, but I preferred to write every line of SHELX myself and did not use these libraries; over the years this has certainly enhanced portability because the programs have not become time-locked into a particular computing environment. The programs are written with (by modern standards) totally excessive attention to optimizing execution speed and the use of memory; a negative side-effect of this is that compiler optimization rarely produces much improvement in performance. Maybe the spartan programming style, for example, the restriction to a few single dimension arrays with one letter names and the terse comments—originally so that the punched cards could be squeezed into one box—has simply deterred 'improvements' to the program code.

The User Interface

An important part of SHELX, and one to which I gave a great deal of thought, is the user interface. The number of input and output files is kept to an absolute minimum and the programs use no configuration files or environment variables. So for a structure refinement, the (usually statically linked) executable SHELXL should be put somewhere in the PATH and two input data files with the extensions .hkl (for the reflection data) and .ins (for everything else) are all that are required. SHELX-76 was often run from a single card-deck by concatenating the condensed data reflection data (see above) onto the end of the remaining data using 'HKLF-1'. If one could find a card-reader, the same card-deck could be fed into SHELXL-97 today and would produce sensible results. Some users have still not forgiven me for the last small change I made to the format of the .hkl reflection data file (in 1975). Since remaining compatible has the highest priority, I could not change this format again now, though it would make a great deal of sense to put the unit-cell that corresponds to the indexing in the file before the first reflection.

The .ins input file was designed to be edited by humans, not computers. Extensive use of *default values* keeps it short. Default values require careful planning because they get used 99% of the time! Free format input was a rarity when SHELX-76 came out; it was not supported by FORTRAN-66 and so had to be encoded in FORTRAN, character by character, but at least this was fully portable. Four-letter words play an important role both in the SHELX input and in the English language in general! In addition to the default values, there is also another feature of the .ins file that makes it very difficult to parse with another computer program; to save space I did not—like the PDB and other formats—start each atom with 'ATOM', so an atom name is simply a keyword that does not have some other defined meaning. Again, it would be nice to change this but retaining upwards compatibility is even more important.

Refinement strategy

Most of SHELX is based primarily on ideas of other people, in particular users of the program. About 90% of my own innovations that I tried to include in the program turned out to be useless; I was careful to eradicate all traces of these so that no one would be tempted to misuse them. The few innovations that turned out to be useful in structure refinement are worth commenting on here. One of these was the introduction of *free variables*, which enabled linear constraints to be applied in a simple and general way; to do this with other programs often required the user to write a special subroutine for each case. Special position constraints were a major application of free variables in SHELX-76, by SHELX-97 the recognition and constraint of special positions had been fully automated, and the most common application of free variables today is probably their use to couple the refinement of the occupancies of different disordered atoms and groups. Many other protein refinement programs still lack special position and occupancy constraints. Rigid group definition (and the removal of rigid group constraints) is very simple and intuitive in SHELXL, though the potentially powerful use of quaternions to fit standard fragments to selected electron density peaks has been widely ignored by users. The use of a connectivity array and PART numbers provides a simple and effective framework for defining disorder and generating hydrogen atoms and various restraints; other macromolecular programs tend to use a much more complicated template approach in which all bonds, hydrogen atoms, etc. are defined in template libraries (this is why some protein graphics programs cannot draw disulfide bonds!). The use of a circular difference Fourier to find the best positions for hydrogen atoms in -OH and -CH_3 groups was another SHELXL-97 innovation. The 'similar distance' restraints and the restraints on the anisotropic displacement parameters (DELU, SIMU and ISOR) also first came into wide use in SHELXL-97, though the rigid bond restraint was probably first used by John Rollett. These restraints are essential both for macromolecular refinement and for handling disorder (often of solvent molecules) in small molecule structures. I am sure that we will be able to find better ways of restraining the displacement parameters in the future this was never intended to be the last word.

I had never imagined that SHELX would eventually find application in macromolecular refinement, and the introduction of several essential features for this purpose can be attributed to encouragement from Zbigniew Dauter and Keith Wilson, who in the early 1990s were looking for ways to refine against the very high resolution protein structures using data collected on the EMBL beamlines at the DESY synchrotron in Hamburg. These features included the solvent model (based on the method used in Dale Tronrud and Lynn Ten Eyck's TNT program) and conjugate gradient solution of the least-squares normal equations (as in John Konnert and Wayne Hendrickson's PROLSQ program). I did introduce some convergence acceleration into this CGLS method by taking into account the shifts in the previous cycle; in fact CGLS should be more widely used for large small molecules, it is very robust. However CGLS does not enable the standard uncertainties in the parameters to be estimated, so a final L.S. cycle—usually with BLOC 1 and DAMP 0 0—is required to obtain these esds for macromolecules. The most complex part of SHELXL to program was probably the derivation of standard uncertainties in all derived parameters taking all correlation terms from the full inverted least-squared matrix into account.

One area, still neglected by macromolecular crystallographers, is the refinement of merohedral and non-merohedral twins. Tests by Garib Murshudov and others have shown that a significant fraction of structures deposited in the PDB are seriously in error because twinning had not been taken into account. My colleague Regine Herbst-Irmer made major contributions to the ways of handling and refining twins with SHELXL-97.

This Book

Over the years I have received many requests from publishers and others to write a book on SHELX, but have always rejected them immediately because it is much more fun to write programs than books. So I am very happy that Peter Müller and his team of authors have at last done it for me, enabling me to continue my hobby of writing crystallographic programs without worrying that it is about time that I explained to the rest of the world how to use them.

George M. Sheldrick
Göttingen
November 2005

Preface

Crystallography has become the most important method of structure determination, and the number of textbooks dealing with the basics of crystallography is legion. This book is not just another one, as it tries to cover the more advanced aspects of crystal structure refinement, which have not been much addressed by textbooks so far. It focuses on practical problems in the everyday life of a crystallographer, dealing with the following topics. After an introduction to the peculiarities of SHELXL—the refinement program on which this book is based—the second chapter provides a brief survey of crystal structure refinement. Chapters three and higher address the various aspects of structure refinement, from the treatment of hydrogen atoms to the assignment of atom types, to disorder, to non-crystallographic symmetry and twinning. The chapter describing protein structure refinement introduces peculiarities of the world of macromolecular crystallography and helps the reader to understand the way SHELXL sees proteins: just as very large small molecules. In addition, the book contains two short chapters about structure validation (one for small molecule structures and one for macromolecules), a topic that is all too frequently neglected. In most chapters, the book gives refinement examples, based on the program SHELXL, describing every problem in detail. It also comes with a CD-ROM that provides all files necessary to reproduce the refinements. In this respect this book is like a tutorial you can attend at home or—if you have a laptop—practically anywhere.

This book should be understood as a complement to rather than a substitute for the SHELX reference manual. Many things mentioned only briefly here are explained in great detail in the manual. A pdf version of the SHELX manual is included on the CD-ROM that accompanies this book, and there should be a printout in every crystallographic facility, as it is the ultimate reference to any SHELX question.

The training of a crystallographer frequently reminds me of that of a Jedi Knight: the practical knowledge only goes from the master's mouth to the apprentice's ear and it can be difficult for the outsider or autodidact to become adept without a local guru's help. Even though the lack of Jedi Knights may be more obvious in our society than the lack of good crystallographers, I think this book will be a helpful tool for many structurally active scientists. Unveiling some secrets of the craft, I hope that *Crystal Structure Refinement* will help to reduce the workload of people like Richard Marsh, Richard Harlow and others who are famous for finding out the errors and mistakes in other crystallographers' publications.

When I started working on this book, I decided that, in order to avoid rewriting any of the existing textbooks, I should assume that the reader knows the fundamentals of crystallography. Therefore this book contains nothing about symmetry, generation of X-rays, diffraction theory and so forth. Examples of excellent introductions to the field of crystallography are listed in the 'Further Reading' section at the back of this book and the reader may turn to one of these sources.

Most chapters were written by me. The chapter on twinning is by Regine Herbst-Irmer, that on protein refinement by Thomas Schneider, the chapter about small-molecule structure validation by Ton Speck, and that on protein structure validation by Michael Sawaya. George Sheldrick, author of SHELXL, wrote the foreword to this book, which includes a brief history of SHELXL.

My warmest thanks go to the four co-authors, who made invaluable contributions to this book. In addition, I am enormously grateful to George Sheldrick for having been my 'Jedi-Master' since 1996 and for his help and support with this project, which would not have been realizable without him. I also wish to express my thanks to Claire Gallou-Müller and Dan Anderson who read all my drafts and supported me in my ideas and whims throughout the writing process.

Peter Müller
Cambridge, MA
December 2005

Contents

Contributors

Dr. Peter Müller
Department of Chemistry
Massachusetts Institute of Technology
77 Massachusetts Avenue, Building 2, Room 325
Cambridge, MA 02139, USA

Dr. Regine Herbst-Irmer
Department of Structural Chemistry
Institute of Inorganic Chemistry
University of Göttingen
Tammannstr. 4
D-37077 Göttingen, Germany

Prof. Dr. Anthony L. Spek
Laboratory of Crystal and Structural Chemistry
Bijvoet Center for Biomolecular Research
Utrecht University
Padualaan 8
3584 CH Utrecht, The Netherlands

Dr. Thomas R. Schneider
IFOM - The FIRC Institute of Molecular Oncology
Biocrystallography and Structural Bioinformatics
Via Adamello 16
I-20139 Milan, Italy

Dr. Michael R. Sawaya
UCLA Technology Center
University of California Los Angeles
Box 951662
Los Angeles, CA 90095-1662, USA

1

SHELXL

This book is about crystal structure refinement with SHELXL, a program written and maintained by George M. Sheldrick. SHELXL is by far the most popular refinement program for small molecules and, together with CNS by Axel Brünger and Refmac by Garib Murshudov, one of the three most commonly used programs for the refinement of protein structures. In the foreword to this book George Sheldrick gives a brief overview of the history of SHELXL and tells how the first versions came to exist.

1.1 The SHELX program suite

SHELXL is part of the SHELX program suite, a software package containing the following programs:

SHELXS	Structure solution by Patterson and classical direct methods.
SHELXD	Structure solution not only for macromolecules.
SHELXL	Structure refinement.
SHELXH	Structure refinement for very large structures (in principal identical with SHELXL, but the maximum number of allowed parameters is larger).
CIFTAB	Tables for publication etc. from the .cif file.
SHELXPRO	This protein interface to SHELX is a collection of different routines to convert file formats, calculate electron density maps, etc.
SHELXWAT	Automatic water divination for macromolecules.

All these programs can be obtained free of charge[1] from George Sheldrick for a multitude of operating systems (www.shelx.uni-ac.gwdg.de/SHELX); the only requirement is to fill out a registration form.

1.1.1 SHELXTL and other programs

There is a commercial twin to the SHELX suite: The SHELXTL package as sold by Bruker AXS[2] contains the programs XS (SHELXS), XM (SHELXD), XL (SHELXL), XH (SHELXH), XCIF (CIFTAB like), XPRO (SHELXPRO), XWAT (SHELXWAT), and some additional programs by George Sheldrick—XP

[1] For the non-profit user, that is. [2] www.bruker-axs.de/products/scd/shelxtl.php

and XPREP—and other authors, like PLATON[3] by Anthony L. Spek or XSHELL. The entire collection of programs can be operated from one central graphical user interface. This interface also opens text editors for the .ins, .res, .lst and .cif files and keeps track of the correct file names.

XPREP is a program for the interactive analysis of diffraction data. Among many other things it assists in the determination of the space group, calculates and displays intensity statistics, generates different kinds of Patterson maps, detects merohedral twinning and sets up the input files for SHELXS and SHELXD (XS and XM in the SHELXTL world). Unfortunately XPREP is not a free program and available only commercially from Bruker-AXS. Nevertheless it is a very helpful tool for any kind of crystallographic work and, while it is not absolutely required to have XPREP in order to work through the examples in this book, we would tend to highly recommend its purchase to anyone doing crystallographic work.

XP is a relatively old but still widely used program to display, analyze and manipulate crystal structures. XP is essentially a graphics application to SHELXL. It reads and writes .ins and .res files and allows examining and manipulating structural coordinates. In addition, several types of figures like Ortep-style thermal-ellipsoid plots, electron density diagrams, etc. can be generated by XP. The most intriguing features are probably the generation of symmetry equivalent atoms and/or molecules and the total freedom in applying any given symmetry to the atoms of a structure.

While XP and XPREP are excellent programs and very helpful, they are not the only tools for the task. Instead of XPREP, one could use for example SORTAV (obtainable as part of the CCP4 program suite[4] or in a stand-alone version directly from the author Robert H. Blesssing). ORTEP-III (available free of charge to academic users from the authors, Michael N. Burnett and Carroll K. Johnson)[5] or ORTEP-3 for Windows (obtainable free of charge by academic users from the author, Louis J. Farrugia)[6] are possible alternatives to XP. To work through the examples in this book it is not necessary to purchase SHELXTL or any other program. Everything described in the example sections can be done with software that is free for the academic or non-profit user.

XSHELL is a graphical interface to SHELXL and goes back to Bob Sparks. It is designed to help with editing the .ins and .res files, picking atom types interactively, and displaying the structure. The program is a direct interface to SHELXL, which means that one can start refinement cycles of SHELXL from XSHELL and have the results displayed immediately after the refinement is finished. XSHELL has some features which have no counterpart in XP, is largely mouse-driven and has a more modern look and feel to it. Unfortunately, Bob Sparks died in 2001 and could not complete his work on XSHELL. At least in the current version, XSHELL is flawed: the handling of restraints, disordered and/or twinned structures, the weighting scheme and many other details make clear that whoever finished writing the

[3] PLATON is available free of charge in a stand alone version from www.xraysoft.chem.uu.nl. More details about this program are given in Chapter 9. [4] www.ccp4.ac.uk/main.html

[5] www.ornl.gov/sci/ortep/ [6] www.chem.gla.ac.uk/~louis/software/ortep3/

program was not a crystallographer. Even though XSHELL is a convenient and easy-to-use program, I strongly recommend you not to use it for anything else than perhaps picking the element types and labelling atoms. It is more than likely that in the near future a new version of XP will offer the same convenience paired with the reliability for which XP is known.

In addition to the programs mentioned, there is a multitude of crystallographic programs, most of them free for the academic user. Some of them (twinrotmap, cell_now, Gemini, SAINT, Coot, XtalView and others) are mentioned in this book and a quick overview is given at the respective places.

Besides SHELXTL, there are other graphical user interfaces to SHELXL and other crystallographic programs. The most popular is probably WinGX by Louis Farrugia. WinGX is an integrated system of publicly available programs for the analysis and refinement of single crystal X-ray diffraction data. It is primarily focused on small molecule crystallography and has been developed out of the Glasgow GX package (hence its name). It can be obtained free of charge from the author.[7]

1.2 SHELXL

Large parts of this section are (almost) literal quotations from the SHELX-97 Manual by George M. Sheldrick, which can also be found on the CD-ROM that accompanies this book. A printout of this manual should be available in every crystallographic facility, as it is the ultimate reference to any SHELX question.

SHELXL is a program for the refinement of crystal structures from diffraction data, and is primarily intended for single crystal X-ray data of small moiety structures, though it can also be used for the refinement of macromolecules against data to about 2.5 Å or better. It uses a conventional structure factor summation, so it is much slower (but a little more accurate) than standard FFT-based macromolecular programs. SHELXL is intended to be easy to install and use. It is general, and is valid for all space groups and types of structure. Polar axis restraints and special position constraints are generated automatically. The program can handle twinning, complex disorder, absolute structure determination, CIF and PDB output, and provides a large variety of restraints and constraints for the control of difficult refinements. The interface program SHELXPRO allows macromolecular refinement results to be displayed in the form of Postscript plots, and generates map and other files for communication with widely used macromolecular programs. The auxiliary program CIFTAB is useful for tabulating the refinement results via the CIF output file for small molecules.

1.2.1 Program organization

Even though several graphical user interfaces to SHELXL have been written, SHELXL is entirely input file based. To run SHELXL only two input files are

[7] www.chem.gla.ac.uk/~louis/software/wingx/

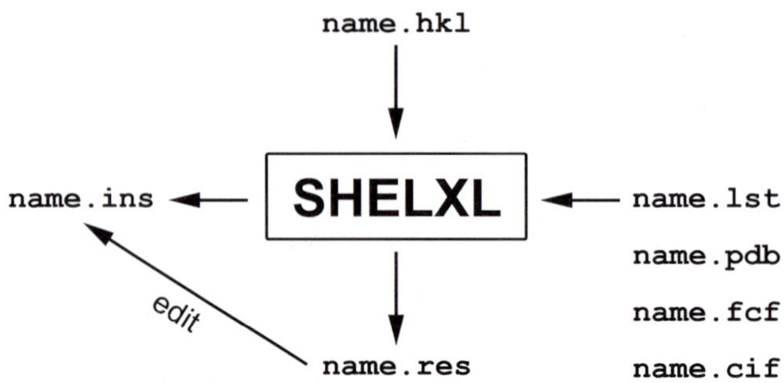

Fig. 1.1 File organization in SHELXL.

required (atoms/instructions and reflection data); since both these files and the output files are pure ASCII text files (and the instructions are strictly not case-sensitive), it is easy to use the program on a heterogeneous network. The reflection data file **`name.hkl`** contains h, k, l, F^2, and $\sigma(F^2)$ in standard SHELX format. The program merges equivalents and eliminates systematic absences, and the order of the reflections in this file is unimportant. Crystal data, refinement instructions and atom coordinates are all input as the file **`name.ins`**. Instructions appear in the .ins file as four-letter keywords followed by atom names, numbers, etc. in free format. There are sensible default values for almost all numerical parameters. SHELXL is normally run on any computer system by means of the command:

`shelxl name`

where `name` defines the first component of the filename for all files which correspond to a particular crystal structure. The executable program must be accessible via the path (or equivalent mechanism). No environment variables or extra files are required.

A brief summary of the progress of the structure refinement appears on the console, and a full listing is written to a file **`name.lst`**, which can be printed or examined with any text editor. After each refinement cycle a file **`name.res`** is (re)written. The .res file is similar to the .ins file, but has updated values for all refined parameters. It may be copied or edited to **`name.ins`** for the next refinement run (Figure 1.1).

1.2.2 The instruction file **`name.ins`**

All instructions commence with a four (or fewer) character word (which may be an atom name); numbers and other information follow in free format, separated by one or more spaces. Upper and lower case input may be freely mixed (with the exception of the text string input using `TITL`); the input is converted to upper case for internal use in SHELXL. The `TITL`, `CELL`, `ZERR`, `LATT` (if required), `SYMM` (if required), `SFAC`, `DISP` (if required) and `UNIT` instructions must be given *in that order*; all remaining instructions, atoms, etc. should come between `UNIT` and the last instruction, which is always `HKLF` (to read in reflection data). A number of instructions allow atom names to be referenced; use of such instructions without any atom names means 'all non-hydrogen atoms'. A list of atom names may also be abbreviated to the first atom, the symbol > (separated by spaces), and then the last atom; this means 'all atoms between and including the two named atoms but excluding hydrogen atoms'.

1.2.3 The reflection data file **`name.hkl`**

The .hkl file consists of one line per reflection in `FORMAT(3I4,2F8.2,I4)` for h, k, l, F_o^2, $\sigma(F_o^2)$, and (optionally) a batch number. This file should be terminated by a record with all items zero; individual data sets within the file should *not* be separated from one another—the batch numbers serve to distinguish among groups of reflections for which separate scale factors are to be refined. The reflection order and the batch number order are unimportant. The .hkl file is read when the `HKLF` instruction (which terminates the .ins file) is encountered. The `HKLF` instruction specifies the format of the .hkl file, and allows scale factors and a reorientation matrix to be applied. Lorentz, polarization and absorption corrections are assumed to have been applied to the data in the .hkl file. Note that there are special extensions to the .hkl format for Laue and powder data, as well as for twinned crystals that cannot be handled by a `TWIN` instruction alone.

1.2.4 Merging data in SHELXL

SHELXL automatically rejects systematically absent reflections. The sorting and merging of the reflection data is controlled by the `MERG` instruction. Usually `MERG 2` (the default) will be suitable for small molecules: equivalent reflections are merged and their indices converted to standard symmetry equivalents, but Friedel opposites are not merged in non-centrosymmetric space groups. `MERG 4`, which merges Friedel opposites and sets $\delta f''$ for all elements to zero, saves time for macromolecules with no significant dispersion effects. F_o^2 means the observed experimental measurement, which may possibly be slightly negative if the background is higher than the peak as a result of statistical fluctuations (see also Chapter 2). The merging residual values R_{int} and R_{sigma} as calculated by SHELXL are defined as follows:

$$R_{\text{int}} = \frac{\sum |F_o^2 - \langle F_o^2 \rangle|}{\sum F_o^2} \tag{1.1}$$

In this equation both summations involve all input reflections for which more than one symmetry equivalent are averaged. $\langle F_o^2 \rangle$ is the mean value of all measured equivalents.

$$R_{\text{sigma}} = \frac{\sum \sigma(F_o^2)}{\sum F_o^2} \tag{1.2}$$

In this equation, the summation occurs over all reflections in the merged list, and $\sigma(F_o^2)$ is the estimated standard uncertainty of the merged reflection. In estimating the σ-value for a merged reflection, SHELXL uses the value obtained by combining the $\sigma(F_o^2)$ values from the individual contributors, unless the estimated standard uncertainty of the mean is larger, in which case it is used instead.

1.2.5 The connectivity table

The key to the automatic generation of hydrogen atoms, molecular geometry tables, restraints and so forth is the connectivity array. This table can be found in the .lst file and contains information on the atoms to which every individual atom in a structure binds. For a non-disordered organic molecule, the connectivity array can be derived automatically using standard atomic radii. A simple notation for disordered groups allows most cases of disorder to be processed with a minimum of user intervention. Each atom is assigned a `PART` number `n`. The usual value of `n` is `0`, but other values are used to label components of a disordered group (see Chapter 5). Bonds are then generated for atoms that are close enough only when either at least one of them has `n = 0`, or both values of `n` are the same. A single shell of symmetry equivalents is automatically included in the connectivity table.

The generation of equivalents (e.g. in a toluene molecule on an inversion centre) may be prevented by assigning a negative `PART` number. If necessary, bonds may be added to or deleted from the connectivity array using the `BIND` or `FREE` instructions. To generate additional bonds to symmetry equivalent atoms, `EQIV` can be used.

2

Crystal structure refinement

The determination of a crystal structure consists of several steps of which refinement comes near-last. In the beginning the crystal needs to be grown and mounted onto the X-ray diffractometer. The next step is the determination of the unit cell and data collection, preferably at low temperature (say 100 K), following a strategy that gives a complete set of data and a high multiplicity of observations (MoO). This is best achieved with a goniometer possessing either three-circle or kappa geometry.[1] Third, in data reduction, a wide and error-prone field, the raw intensities from the detector are translated into structure factors (or in most cases squared structure factors). In this step, the data reduction programs apply several corrections (such as Lorentz, polarization, absorption, etc.) and determine values for the standard uncertainties for each reflection (sigmas). Numerous methods are employed to determine the phase angle for each of the structure factors, the fourth stage of the crystal structure determination, also called structure solution. The choice of the best of these methods depends on the individual problem (size of the structure, presence of heavy atoms, presence of anomalous scattering, maximum resolution, and so forth).

After these four steps the crystallographer has obtained atomic coordinates for some or all non-hydrogen atoms. Frequently the atom types assigned to some of these coordinates are incorrect or no atom types have been assigned at all. In addition, the coordinates in the first solution are usually not very accurate: they might be within 0.1 Å of the correct positions. Moreover, many details of the structure are yet to be determined: groups of lighter atoms, disorders, hydrogen positions, etc. The way from the first solution to the final accurate and publishable model is called refinement. Depending on the structure, this can be a short highway to happiness or a long and rugged road through pain and sorrow.

Oftentimes the short highways can be travelled on auto pilot, and ‘carriage return’ becomes the most important, if not the only, key on the computer. Referring to the decimal value of the carriage return key in the ASCII character set, we could call this the Highway 13—and that is not what this book is about. This book is about the outdoor adventure of roaming the rough roads of refinement, those perilous paths of

[1] From the data-quality point of view, the kappa geometry, which allows for three independent axes of crystal rotation, is somewhat better than the three-circle, which offers only two independent crystal rotations (ω and ϕ, while the χ angle is fixed to 54.7°). A true four-circle diffractometer is more versatile than a three-circle and comparable to the kappa geometry with respect of reciprocal-space coverage. However the bulky and relatively heavy Eulerian cradle of the four-circle restricts the effective ω-range due to shadowing and prevents or complicates installation of a low temperature device, video camera, etc. The advantage of three-circle over kappa geometry is its robustness and lower price.

ponderous progress, full of problems and pitfalls. Yet before we get carried away in alliterations, let us talk about refinement.

2.1 Least-squares refinement

The atomic positions in the first solution are not the direct result of the diffraction experiment but an interpretation of the electron density calculated from the measured intensities and the 'somehow-determined' initial phase angles. New, usually more accurate phase angles can be calculated from the atomic positions, which allow re-determining the electron density function with a higher precision. From the updated electron density map, more accurate atomic positions can be derived, which lead to even better phase angles, and so forth. New atoms can be introduced into the model, when the most recent electron density function shows a high value at a place in the unit cell where the model does not contain an atom yet. Sometimes, atoms need to be removed from the model when they occupy positions in the cell corresponding to a low value in the electron density function. When the atomic model is complete, atoms can be described as ellipsoids rather than spheres (anisotropic refinement) and hydrogen atom positions can be determined or calculated. Every step in this process is undertaken to improve the accuracy of the model, and the entire procedure from the initial atomic positions to the complete, accurate and (if achievable) anisotropic model with hydrogen positions is called the refinement.

A critical point in this process is the evaluation of the model, as the model should only be altered if a change improves its quality. There are several mathematical approaches to define a function which is assumed to possess a minimum for the best possible model: in the world of small molecules (typically less than 200 independent atoms) the least-squares approach is by far the most common method, while for protein structures other methods like maximum likelihood have also been employed. The program SHELXL, on which this book focuses, is predominantly a program for small-molecule structures and the least-squares refinement is the only method on which we need to concentrate.[2] The concept is simple: by means of Fourier transformation, a complete set of structure factors is calculated from the atomic model. The calculated intensities are then compared with the measured intensities, and the best model is that which minimizes M:

$$M = \sum w \left(F_{\text{o}}^2 - F_{\text{c}}^2\right)^2 \tag{2.1}$$

or

$$M = \sum w \Big(|F_{\text{o}}| - |F_{\text{c}}|\Big)^2 \tag{2.2}$$

[2] SHELXL uses the least-squares method for the refinement of small and macromolecular structures alike, even though there is a choice between full-matrix least-squares and a conjugate gradient version. The first is more accurate and matrix inversion gives access to standard uncertainties for all distances and angles, etc. Conjugate gradient least-squares on the other hand is much faster, which makes it more appropriate for protein refinements and the early stages of the refinement of larger small-molecule structures.

In these two functions, F is the structure factor and the subscripts o and c stand for observed and calculated; this nomenclature is going to be used throughout the book. Each addend in this summation is multiplied by an individual weighting factor w, which reflects our confidence in this particular datum and is derived from the standard uncertainty σ of that measurement.[3] The only difference between these two minimization functions is that the left one corresponds to refinement against squared structure factors (F^2), while the right equation describes refinement against F-values.

2.1.1 *Refinement against* F *or* F^2*—is that a question?*

In the past, refinement was usually performed against structure factors F. In order to minimize the above function (Equation 2.2), the measured intensities have to be transformed into structure factors. This involves the extraction of a root (bear in mind: $I \propto F^2$), which leads to mathematical problems with very weak reflections or reflections with negative measured intensities.[4] To circumvent this problem, negative measurements must be set to zero, or to an arbitrary small positive value in a refinement against F. Such an approach introduces bias, as the very weak reflections do contain information and ignoring them affects the structure determination. Another problem in the use of F values arises from the difficulty of estimating the $\sigma(F)$ values from the $\sigma(F^2)$ values, the latter of which are determined during data reduction. As the least-squares method is very sensitive to the weights applied to each reflection in the above summation, problems with the σ estimation lead to inaccuracies in the refinement.

Refinement against F^2 (Equation 2.1) does not cause any of these problems and even has additional advantages: it makes the refinement of twinned structures mathematically simpler, and refinement against squared structure factors is less likely to settle into a local minimum. Therefore, refinement against F^2 is superior to refinement against F, even though some more traditional crystallographers still insist on refining against structure factors.

A broader discussion of this matter is beyond the scope of this book, but the interested reader may turn to the articles by Hirshfeld and Rabinovich (1973) and Arnberg *et al.* (1979) for more in-depth information.

2.2 Weak data and high-resolution cut-off

As mentioned above, it is important not to exclude weak data. However, there is no reason to use data from high-resolution shells when they are all very weak, since these reflections are in fact noise and contain no usable information. Generally, intensities are weaker at higher 2Θ angles and almost no crystal diffracts to the theoretical limit of $d_{\max} = \lambda/2$. Some care must be taken in the determination of the effective maximum resolution of a dataset.

[3] Frequently: $w = 1/\sigma$

[4] Because of counting statistics (background higher than signal of the peak), sometimes slightly negative intensities are measured for very weak reflections.

There are two major criteria to be taken into consideration: the $\langle I/\sigma \rangle$ as a measure of the strength of the signal, and the merging R value R_{int} (or R_{sigma}) of all data within a shell. Generally, the $\langle I/\sigma \rangle$ values become smaller with higher resolution, while the R_{int} values grow. What now are the minimum values for $\langle I/\sigma \rangle$ and the maximum values for R_{int} that distinguish 'data' from 'noise'? This question is not answered easily, but many crystallographers agree that data with overall values of $\langle I/\sigma \rangle \leq 2.0$ and/or $R_{\text{int}} \geq 0.45$ throughout a certain resolution shell are to be considered noise. In practice there are more factors to be taken into account and, as always, experience helps. The following table represents the statistics of a dataset collected with a CCD detector. The edge of the detector corresponds to a resolution of 0.77 Å, but the crystal did not diffract quite that far. In the table, the dataset is subdivided into resolution shells, and the Completeness, Multiplicity of Observation (MoO),[5] the mean intensity over the standard uncertainty ($\langle I/\sigma \rangle$) and two different merging R values (R_{int} and R_{sigma})[6] are given.

Resolution	Compl.	MoO	$\langle I/\sigma \rangle$	R (int)	R (sigma)
Inf. – 2.15	99.2	9.27	43.21	0.0294	0.0171
2.15 – 1.70	100.0	9.49	31.76	0.0455	0.0210
1.70 – 1.50	100.0	7.86	28.57	0.0450	0.0242
1.50 – 1.35	100.0	6.95	22.07	0.0588	0.0316
1.35 – 1.25	100.0	6.33	18.28	0.0761	0.0395
1.25 – 1.15	100.0	5.72	14.60	0.0960	0.0511
1.15 – 1.05	100.0	5.18	11.33	0.1365	0.0712
1.05 – 1.00	100.0	4.67	8.49	0.1848	0.0992
1.00 – 0.95	99.7	4.22	7.53	0.2066	0.1193
0.95 – 0.90	98.8	3.79	5.22	0.2873	0.1774
0.90 – 0.85	94.3	3.08	3.75	0.3928	0.2632
0.85 – 0.80	61.1	0.87	1.94	0.4933	0.4777
0.80 – 0.77	16.9	0.17	1.50	0.4704	0.5981
0.90 – 0.77	58.2	1.34	2.79	0.4062	0.3760
Inf. – 0.77	84.2	4.19	12.90	0.0715	0.0653

When comparing the various entries of this table throughout the resolution shells, it becomes obvious that the inner data are complete, very strong (high $\langle I/\sigma \rangle$), and fairly accurate (low merging R values). With increasing resolution, the data become weaker and somewhat less accurate, which is a normal trend, but the completeness does not change significantly. At resolutions higher than 0.85 Å, however, the

[5] 'This term was defined at the SHELX workshop in Göttingen in September 2003 to distinguish the MoO from redundancy, or multiplicity, with which the MoO has frequently been confused in the past. In contrast to redundancy, which is repeated recording of the same reflection obtained from the same crystal orientation (performing scans that rotate the crystal by more than 360°), MoO, sometimes also referred to as "true redundancy", describes multiple measurements of the same reflection obtained from different crystal orientations (i.e. measured at different ψ–angles)'. Quoted from Müller *et al.* (2005).

[6] Definitions of the merging R-values can be found in Chapter 1.

completeness drops to very low values. In addition, the $\langle I/\sigma \rangle$ values suggest that the outer data are mostly noise and the merging R values are well above the threshold for acceptable data. The overall statistics for the whole dataset (last line of the table) show very good numbers for $\langle I/\sigma \rangle$ and quite acceptable merging R values. The completeness and multiplicity of observations (MoO), however, are rather poor. The dataset corresponding to the above data statistics should be truncated at 0.85 Å, or maybe even at 0.90 Å. When in doubt we should go for the more conservative option and apply a high-resolution cut-off at 0.85 Å. The new data statistics look as follows:

Resolution	Compl.	MoO	$\langle I/\sigma \rangle$	R (int)	R (sigma)
Inf. – 2.30	99.0	9.11	44.10	0.0287	0.0171
2.30 – 1.80	100.0	9.88	33.20	0.0436	0.0199
1.80 – 1.55	100.0	8.24	30.54	0.0426	0.0226
1.55 – 1.40	100.0	7.23	25.00	0.0515	0.0279
1.40 – 1.30	100.0	6.58	19.04	0.0727	0.0376
1.30 – 1.20	100.0	6.02	16.13	0.0884	0.0461
1.20 – 1.15	100.0	5.59	13.95	0.0962	0.0520
1.15 – 1.10	100.0	5.35	12.90	0.1131	0.0605
1.10 – 1.05	100.0	5.04	10.06	0.1642	0.0829
1.05 – 1.00	100.0	4.67	8.49	0.1848	0.0992
1.00 – 0.95	99.7	4.22	7.53	0.2066	0.1193
0.95 – 0.90	98.8	3.79	5.22	0.2873	0.1774
0.90 – 0.85	94.3	3.08	3.75	0.3928	0.2632
0.95 – 0.85	96.3	3.40	4.43	0.3335	0.2184
Inf. – 0.85	98.9	5.46	14.58	0.0703	0.0508

Now the statistics for the whole dataset (last line) have much improved: an overall completeness of 99% is fine and a MoO of 5.5 is acceptable. These are the data against which we should refine the model.

2.3 Residual factors

The quality of the model can be judged with the help of various residual factors or 'R-factors'. These factors should converge to a minimum during the refinement and are to be quoted when a structure is published. The three most commonly used residual factors are:

The weighted R-factor based on F^2: wR (or $wR2$ in SHELXL), which is most closely related to the refinement against squared structure factors.

$$wR = \left[\frac{\sum w \left(F_o^2 - F_c^2\right)^2}{\sum w F_o^2} \right]^{1/2} \tag{2.3}$$

The weighting factor w is individually derived from the standard uncertainties of the measured reflections and expresses the confidence we have in every single reflection.

Albeit based on F values and hence mostly of historical value, the most popular one is the unweighted residual factor based on F: R (or $R1$ in SHELXL).

$$R = \frac{\sum \left||F_o| - |F_c|\right|}{\sum |F_o|} \tag{2.4}$$

Finally, there is the goodness of fit: GooF, GoF, or simply S.

$$S = \left[\frac{\sum w\left(F_o^2 - F_c^2\right)^2}{(N_R - N_P)}\right]^{1/2} \tag{2.5}$$

In this equation N_R is the number of independent reflections and N_P the number of refined parameters. Theoretically, for a properly adjusted weighting scheme, the value for S should be close to 1. However, manipulating or rescaling the weights w can artificially improve this value. In fact, SHELXL uses the above formula to calculate a suggested weighting scheme. This makes it important not to adjust the weights too early in the refinement—by no means before all atoms have been included into the model—as the number of parameters influences the value of S and hence the weighting scheme suggested by SHELXL.

A goodness of fit of $S < 1$ suggests the model is better than the data. Obviously this is suspicious and usually a sign that there are some problems with the data and/or the refinement. Frequently, failure to perform a proper absorption correction leads to underestimated GooF values, but refinement in the wrong space group can also have this effect.

For macromolecular structures, there is one additional residual factor, the R_{free}, introduced by Axel Brünger (1992) which provides a tool to detect overfitting (see also Chapters 10 and 11).

2.4 Parameters

For every atom in the model that is located on a general position in the unit cell, there are three atomic coordinates and one or six atomic displacement parameters (one for isotropic, six for anisotropic models) to be refined. In addition there is one overall scale factor per structure (*osf*, or the first free variable in SHELXL; see Section 2.7) and possibly several additional scale factors, like the batch scale factors in the refinement of twinned structures, the Flack-x parameter for non-centrosymmetric structures, one parameter for extinction, etc. In addition to the overall scale factor, SHELXL allows for up to 98 additional free variables to be refined independently. These variables can be tied to site occupancy factors (see Chapter 5) and a variety of other parameters such as interatomic distances.

Besides these, there is a second group of parameters, which also have significant influence on the structure: the atom types. Even though the atom types are not refined but set by the crystallographer, they determine which atomic scattering factors are used in the Fourier transformation. An incorrectly assigned atom type can cause several kinds of problems, such as making the refined parameters related to this atom adjust to incorrect, and sometimes even meaningless values in an attempt to compensate for the wrong atom type.

It can be deduced from the above that the overall number of parameters depends mostly on the number of crystallographically independent atoms and can be assumed to be roughly 9–10 times the number of atoms in the asymmetric unit for an anisotropic model (see also Figure 10.1). A stable and reliable refinement requires a minimum number of observations per refined parameter, and the International Union of Crystallography (IUCr) currently recommends a minimum data-to-parameter ratio of 8 for non-centrosymmetric structures and 10 for centrosymmetric structures. This corresponds to a resolution of about 0.84 Å or a $2\Theta_{\mathrm{max}}$ of 50° for Mo $K\alpha$ radiation and 134° for Cu $K\alpha$ respectively.[7] In many small molecule cases it is not difficult to collect data to 0.75 Å or better, but sometimes a crystal does not diffract well enough. In such cases constraints and more importantly restraints can help to indirectly improve the data-to-parameter ratio.

2.5 Constraints

Constraints are equations rigidly relating two or more parameters or assigning fixed numerical values to certain parameters, hence reducing the number of independent parameters to be refined. The following paragraphs give an overview of constraints commonly used in crystal structure refinement. An excellent description of the use of constraints and restraints in crystal structure refinements has been given by Watkin (1994).

2.5.1 *Site occupancy factors*

One constraint found in practically every refinement is the site occupancy factor. In the absence of disorder it is fixed to unity, which means that the atom site is fully occupied (in other words the atom is present at that site in every unit cell). For atoms disordered over two sites in the unit cell, the ratio of the two site occupancy factors can be refined, but generally their sum is still constrained to unity.

2.5.2 *Special position constraints*

Atoms on special positions require constraints for their coordinates and sometimes also their anisotropic displacement parameters. In addition the occupancies of atoms on special positions—and sometimes also of those atoms bound to them—need to

[7] This is assuming that all symmetry-equivalent reflections are merged (except Friedel pairs for non-centrosymmetric structures) and not treated as independent for the calculation of the data–parameter ratio.

reflect the multiplicity of the special positions. These constraints are called special position constraints, and Table 2.1 gives a few examples.

Figure 2.1 shows a cartoon of an atom on a twofold axis along b and the consequences for coordinates and anisotropic displacement parameters.

2.5.3 *Rigid group constraints*

A rigid group is a number of atoms in a given spatial arrangement, for example five atoms on a perfect pentagon (Cp ligand) or five atoms forming a SO_4 tetrahedron.

Table 2.1 Examples of special position constraints on coordinates, anisotropic displacement parameters and site occupancy factors

Special position	Constraints on coordinates	Constraints on U^{ij} values	Constraints on occupancies
Inversion Centre	x, y, z fixed to lie on inversion centre	None	0.5
Mirror plane $\perp$ to y	y is fixed to lie on the mirror plane	$U^{12} = U^{23}$	0.5
Twofold axis parallel to y	x and z fixed to lie on the twofold axis	$U^{23} = U^{12} = 0$	0.5
Tetragonal fourfold	y and y fixed to lie on the fourfold axis	$U^{11} = U^{22}$ and $U^{12} = U^{13} = U^{23} = 0$	0.25

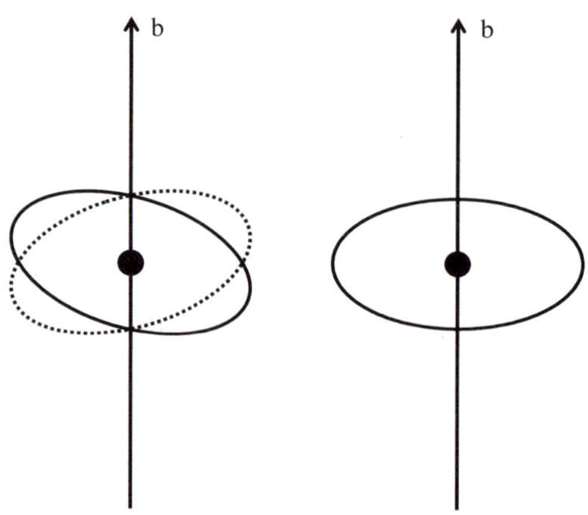

Fig. 2.1 Atom on a twofold axis along y. A 180° rotation must not change the position of the atom or the shape of the thermal ellipsoid. From the first condition follows: $(x, y, z) = (-x, y, -z)$, which is only true for $x = z = 0$. The second condition dictates: $(U^{11}, U^{22}, U^{33}, U^{23}, U^{13}, U^{12}) = (U^{11}, U^{22}, U^{33}, -U^{23}, U^{13}, -U^{12})$, which is only true for $U^{23} = U^{12} = 0$. The left-hand side of the figure shows an incorrectly shaped thermal ellipsoid mapped onto itself by the dyad, the right-hand side a correctly shaped one.

In some cases of heavily disordered groups, in which the individual atomic sites cannot be resolved, but the geometry of the group is known (e.g. a sulfate or perchlorate ion), it can be useful to refine the six parameters of a rigid group (three translational and three rotational parameters) instead of the $3N$ required parameters for the individual atoms. In addition to the six parameters mentioned, a seventh one can be refined as a bond-length scaling factor to allow the rigid group to 'breathe'. That scaling factor allows the distances among the atoms in the rigid group to be refined, while retaining the overall geometry.

2.5.4 *Floating origin constraints*

In polar space groups, which have a floating origin (e.g. $P1$ where the origin is entirely arbitrary, or $P2_1$ where the origin can be anywhere on the b axis), a shift of the entire atomic model along a polar axis does not violate the space group symmetry. In the presence of a heavy atom in the structure, this atom's coordinates can be constrained to certain values (e.g. (0, 0, 0) for $P1$ or $(x,\ 0,\ z)$ for $P2_1$). Alternatively, the sum of the coordinates of all atoms in the structure can be constrained to remain constant, which removes one parameter per polar axis.

SHELXL uses a mathematically different and somewhat more stable approach, *restraining* the weighted sum over all coordinates to remain constant as introduced by Flack and Schwarzenbach (1988). A relatively high weight for this floating origin restraint makes it almost equivalent to a constraint.

2.5.5 *Hydrogen atoms*

Hydrogen atoms are frequently placed on geometrically calculated positions and then refined using a 'riding model.' This means applying constraints to the X—H bond lengths and H—X—H or H—X—Y angles, setting them to certain values. If the atom that carries the hydrogen moves about the unit cell, the hydrogen atoms move with it (like a rider moves with the horse), keeping the hydrogen bond lengths and angles constant. These constraints are a variation of the rigid group constraints, treating the hydrogen atoms bound to a non-hydrogen atom as a rigid group, where the parameters of translation, and in most cases also of rotation, are not refined but derived from the coordinates and geometry of the non-hydrogen atom. This is why adding hydrogen atoms to a model does not necessarily increase the number of parameters.

2.5.6 *Constraints in SHELXL*

SHELXL sets most constraints automatically (they can be changed by the user, of course); however some constraints must be applied manually as needed.

First there are the already mentioned rigid group constraints. Rigid groups are defined and constrained with the `AFIX` instruction, which, in combination with two numbers, `m` and `n`, describes the geometry and mathematical treatment of the group. The AFIX command is followed by a list of the atoms in the rigid group and terminated by AFIX 0.

For hydrogen atoms, the HFIX command followed by m and n generates the appropriate AFIX commands together with the hydrogen atoms. A more complete description of the available m and n qualifiers is given in Chapter 3.

Finally, there are two more constraints: EXYZ followed by two atom names, which constrains the atoms named to share identical coordinates, and EADP followed by two atom names, which constrains the two atoms named to have identical anisotropic displacement parameters.

2.6 Restraints

In general, the only assumption made during a refinement is that the structure consists of atoms. It is, however, possible to include all kinds of additional information a chemist or physicist may have about a certain molecule (e.g. that aromatic systems tend to be flat or that the three methyl groups in a tert-butyl moiety are equivalent). This is done with the help of restraints. Restraints are treated as additional experimental observations, hence indirectly increasing the number of data points to refine against. In the presence of restraints the minimization function (Equation 2.1) changes as follows:

$$M = \sum w\left(F_o^2 - F_c^2\right)^2 + \sum 1/\sigma^2(R_t - R_o)^2 \tag{2.6}$$

In this equation σ is the standard uncertainty (or elasticity) assigned to a restraint; R_t is the target value and R_o the actual value of the restrained quantity.

In many refinements, restraints may not be needed at all. However, when the data to parameter ratio is poor, or when correlations among certain parameters occur (e.g. for the refinement of disorders and pseudo-symmetry), restraints can become essential. In general, when using restraints in SHELXL, the list of 'most disagreeable restraints', found in the .lst file, needs to be examined carefully. In this list the target values of the restraints are compared with those resulting from the refinement. In the case of a strong deviation, which indicates that a restraint has been overruled by the diffraction data, the validity of this restraint needs to be verified. If appropriate, the standard uncertainty assigned to that restraint can be decreased, which in turn gives the restraint a greater weight.

Restraints must be applied with great care and only if justified (George Sheldrick said: 'with the right restraints, you can fit an elephant to any data'). When appropriate, however, they should be used without hesitation, and having more restraints than parameters in a refinement is nothing to be ashamed of.

In SHELXL, restraints are applied by adding a command with appropriate keywords and/or atom names in the .ins file. Even though the SHELXL reference manual, which is included as a .pdf file on the CD-ROM that accompanies this book, exhaustively elaborates on all restraints, the following pages briefly describe the most common restraints as they are used by this program.

2.6.1 *Geometrical restraints*

Besides a restraint on chiral volumes (`CHIV`) and a restraint for atoms that are supposed to lie on a common plane (`FLAT`),[8] SHELXL has two kinds of distance restraints: direct and relative distance restraints. The former restrain distances to a given target value (`DFIX`, `DANG`), while the latter restrain equivalent distances to be equal (`SADI`, `SAME`). Relative distance restraints have the advantage that they do not require a target value, which minimizes the amount of 'outside' information imposed on the model. These restraints also improve the convergence of the refinement, especially when the asymmetric unit contains several equivalent molecules. On the downside, relative distance restraints frequently lead to underestimated standard uncertainties of bond lengths and angles. In addition these restraints make it relatively easy to refine a molecule in a space group of lower symmetry, which can lead to the crystallographer being 'Marshed'.[9]

DFIX and DANG

With the help of the distance restraints `DFIX` and `DANG`, the distance between two atoms can be restrained to possess any target value. The syntax for bond distance restraints is

```
DFIX s d atomnames
```

where `s` is the standard uncertainty and `d` the target distance between the first two atoms named in the list of atom pairs, the third and fourth named atoms (if present), and so forth. If `s` is not specified the default value of 0.02 Å is assumed. If, for example the atoms C(1) and C(2) are supposed to have a distance of 1.54 Å between them, the appropriate command would be: `DFIX 1.54 C1 C2`

The `DANG` instruction is used to restrain bond angles, which correspond to 1,3-distances. The only difference between `DFIX` and `DANG` lies in their default standard uncertainty (0.02 Å for `DFIX` and 0.04 Å for `DANG`), which makes `DFIX` more suitable for 1,2-distances. `DANG` is appropriate for 1,3-distances, which can be assumed to be somewhat less rigid. In any case, the default standard uncertainty can be over-ridden manually, and the two command lines

```
DFIX 1.54 C1 C2
```

and

```
DANG 0.02 1.54 C1 C2
```

are identical.

[8] Actually, `FLAT` and `CHIV` both use the same algorithm. `CHIV` restrains the chiral volume of only one atom to any given value and `FLAT` restrains the (chiral) volume of a number of tetrahedra involving the atoms in question to zero.

[9] To be Marshed refers to Richard E. Marsh's long history of exposing structures published in the wrong space group. Together with Richard L. Harlow's 'ORTEP of the Year' award, Marsh's work has encouraged careful work and crystallographic craftsmanship for decades.

SADI

The similarity restraint `SADI` restrains the distance between two (or more) pairs of atoms to be equal within a default standard uncertainty `s` of 0.02 Å, no matter what that distance might be. The syntax is similar to `DFIX` / `DANG`, except that no target value for the distances is specified:

```
SADI s atomnames
```

If you have a reason to assume that the distance between the atoms C(1) and C(2) should be similar to the distance between the atoms C(7) and C(8), the appropriate command line would read:

```
SADI C1 C2 C7 C8
```

Again, the default standard uncertainty of 0.02 Å can be changed by introducing its new value as `s` between `SADI` and the first atom. Notice that this restraint requires atom *pairs*; hence the number of atom names needs to be even.

SAME

With the help of `SAME`, the geometry of two or more groups of atoms can be restrained to be similar. This can be convenient when a structure contains several crystallographically independent but geometrically equivalent molecules or ligands. The `SAME` command generates the necessary `SADI` restraints with appropriate standard uncertainties (0.02 Å for 1,2-distances and 0.04 Å for 1,3-distances) for equivalent molecules or parts of molecules. It is a very powerful restraint but particularly error-prone, since it requires both the atoms named with the restraint and the atoms following the `SAME` command in the .ins file to be in the correct order. The syntax as well as the benefits and pitfalls of `SAME` are explained in Chapter 5.

FLAT

If four or more atoms are supposed to lie on a common plane (e.g. atoms of an aromatic system) one can use `FLAT` to restrain them to do so within a given standard uncertainty `s` (default value 0.1 Å^3). The correct syntax is:

```
FLAT s atomnames
```

To restrain the six atoms C(1) to C(6) of a phenyl ring so that all lie within a plane, the FLAT restraint would look like this:

```
FLAT C1 C2 C3 C4 C5 C6
```

Or, if the six atoms are immediately following each other in the .ins file:

```
FLAT C1 > C6
```

CHIV

The `CHIV` command is a restraint on the chiral volume of an atom. SHELXL defines the chiral volume as the volume of the tetrahedron formed by the three bonds to an atom, which must be bonded to three and only three non-hydrogen atoms in the connectivity list. The sign of the chiral volume is defined by the alphabetical order of the atoms forming the three bonds. The syntax for this restraint is

```
CHIV V s atomnames
```

The default target value of the chiral volume `V` is 0 (restraining an atom's environment to be planar), and the default value of the standard uncertainty `s` is 0.1 Å^3. This restraint is especially useful for the refinement of biological macromolecules (the chiral volume of the alpha carbon atom in an amino acid residue is about 2.5 Å^3).

2.6.2 Restraints on displacement parameters

Two of the three restraints on displacement parameters (`DELU`, `SIMU`) take into account the fact that atoms, which are bound to one another, move similarly, both in direction and amount (Hirschfeld 1976; Didisheim and Schwarzenbach 1987). The third one (`ISOR`) encourages approximate isotropic behaviour for otherwise anisotropically refined atoms. Both `SIMU` and `DELU` are based on physically very sensible assumptions and can be used on all or almost all atoms in a model when the data to parameter ratio is low or other problems with the refinement make this seem desirable. `SIMU`, however, should not be applied to very small ions, isolated atoms and atoms that are part of freely rotating groups.

DELU
This 'rigid bond restraint' is applied to all bonds connecting atoms on the same `DELU` instruction. It restrains the anisotropic displacement parameters of two atoms in the direction of the bond between them to be equal within a given standard uncertainty `s1` (default value 0.01 Å^2). When appropriate, the same restraint is applied to 1,3-distances, employing the standard uncertainty `s2` (default value is also 0.01 Å^2). The syntax is

```
DELU s1 s2 atomnames
```

If no atom names are given, all non-hydrogen atoms are assumed; if `s1` but not `s2` is specified, `s2` is assumed to possess the same value as `s1`. The use of `DELU` is explained and illustrated in detail in Chapter 5.

SIMU
It can be assumed that atoms that are bound to one another would move in similar directions with approximately similar amplitudes. With the syntax

```
SIMU s st dmax atomnames
```

atoms closer to one another than `dmax` (default value 1.7 Å) are restrained to have the same U^{ij} components within the standard uncertainty `s` (default value 0.04 Å^2). For terminal atoms `st` (default value 0.08 Å^2) is used instead of `s`. If no atom names are given, all non-hydrogen atoms are assumed; if `s` but not `st` is specified, `st` is assumed to possess twice the value of `s`. The use of `SIMU` is also explained and illustrated in Chapter 5.

SIMU implies much bolder assumptions than DELU (hence the fourfold higher default standard uncertainty, which gives this restraint a much lower weight) and should not be used for very small molecules and ions, especially when free rotation is possible (C_5H_5-groups or AsF_6-ions). In general, however, both SIMU and DELU are good ways to indirectly improve the data to parameter ratio for larger structures with poor resolution.

ISOR

ISOR restrains the U^{ij} components of anisotropically refined atoms to behave approximately isotropically within a standard uncertainty of s, or st for terminal atoms (default values 0.1 Å^2 and 0.2 Å^2). The syntax is as follows:

```
ISOR s st atomnames
```

If no atom names are given, all non-hydrogen atoms are assumed; if s but not st is specified, st is assumed to possess twice the value of s. The use of ISOR is also explained and illustrated in Chapter 5.

ISOR can be usefully employed for the refinement of solvent water molecules, for which SIMU and DELU are ineffective (Figure 2.2). ISOR can also be used (and very easily abused) to keep certain atoms from becoming non-positive definite.[10] In general, ISOR should always be applied as a weak restraint with relatively large standard uncertainties.

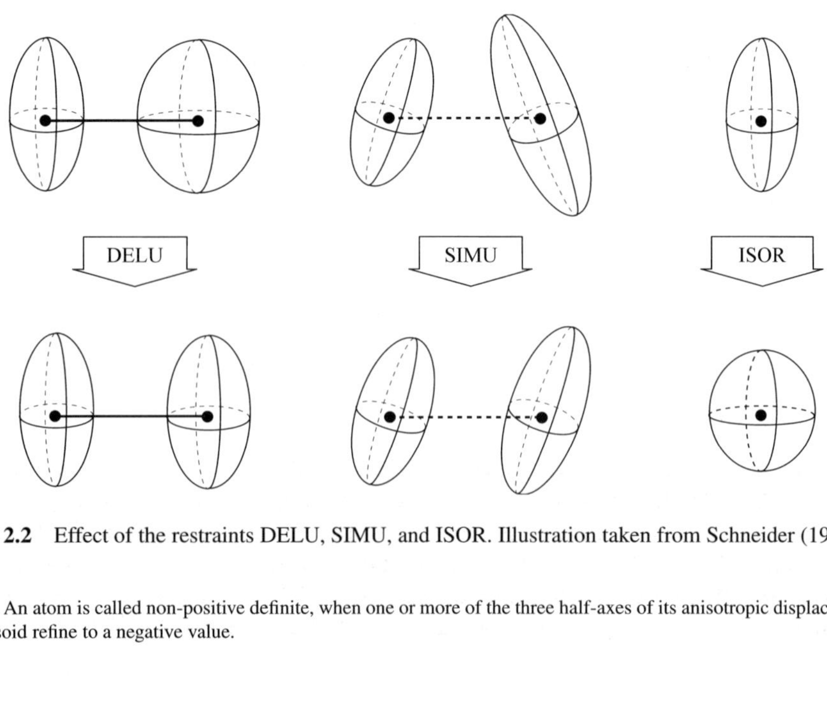

Fig. 2.2 Effect of the restraints DELU, SIMU, and ISOR. Illustration taken from Schneider (1996b).

[10] An atom is called non-positive definite, when one or more of the three half-axes of its anisotropic displacement ellipsoid refine to a negative value.

2.6.3 Other restraints

BUMP

If two atoms that are not connected according to the connectivity table approach each other to a distance shorter than the expected shortest non-bonding distance, BUMP generates a restraint that pushes those atoms apart. Those 'anti-bumping' restraints are generated for the element types C, N, O and S and also for hydrogen atoms that are not bonded to the same atom. This restraint is used almost exclusively for the refinement of macromolecular structures and can help to avoid energetically unfavourable side-chain conformations. It also helps to generate a solvent model with acceptable hydrogen bonding distances that is consistent with the diffraction data. The syntax for BUMP is as follows:

```
BUMP s
```

Here s is the standard uncertainty (default value is 0.02 or the first DEFS parameter). If s is negative, the absolute value is used as standard uncertainty and symmetry equivalent atoms are taken into account when deciding which atoms are connected (that can be interesting when the asymmetric unit contains fractions of a full molecule and bonds go through symmetry elements).

SUMP

The SUMP command allows to restrain the (weighted) sum of several free variables to assume a given target value (see Section 2.7 about free variables). This command can be used to refine disorders with more than two components (for details see Chapter 5) but is not limited to this application. Using the syntax

```
SUMP c sigma c1 m1 c2 m2 ...
```

the following linear equation is applied to the specified free variables:

```
c = c1 · fv(m1) + c2 · fv(m2) + ...
```

where c is the target value for the restraint and sigma the standard uncertainty. c1, c2, etc. are weighting factors and frequently 1; m1, m2, etc. refer to the values of the individual free variables.

DEFS

DEFS globally changes the default standard uncertainties for the restraints CHIV, DANG, DELU, DFIX, FLAT, SADI, SAME and SIMU, using the following syntax:

```
DEFS sd[0.02] sf[0.1] su[0.01] ss[0.04] maxsof[1]
```

In parentheses are the default values. sd is the default for s in DFIX and SADI, and for s1 in the SAME instruction; for DANG twice the value of sd is applied. sf is the default standard uncertainty for CHIV and FLAT, su is the default values for s1 and s2 in DELU, and ss is the default value for s in SIMU. As mentioned above, the default values for st in SIMU and ISOR, as well as s2 in SAME are calculated from the respective s or s1 values (unless specified differently by the user). maxsof specifies the maximum value up to which a site occupancy factor is allowed to refine. Fixed site occupancy factors and *sof*s linked to free variables are not restricted by maxsof. As described above, the standard uncertainties of any given restraint can be defined locally within the individual command. This does not affect the default standard uncertainties or the global standard uncertainties as defined by DEFS for the other restraints.

2.7 Free variables in SHELXL

As their name suggests, free variables can be used to refine a multitude of different parameters and facilitate the formulation of constraints and restraints. The first free variable is always the overall scale factor (*osf*), which is used to bring the reflections in the dataset to an absolute scale. The example in Section 4.4.3 shows the effects of incorrect scaling on the refinement. Additional free variables can be linked to the site occupancy factors (*sof*) of groups of disordered atoms (for details see Chapter 5), but can also be related to other atomic parameters (x, y, z, *sof*, U, etc.) and even interatomic distances, chiral volumes, and other parameters.

In general, any parameter P or any DFIX, DANG, or CHIV restraint can be defined in the .ins file as $10 \cdot m + p$. There are four different cases:

$m = 0$: the parameter P with the starting value p is refined freely.
$m = 1$: the value of p is fixed and not refined at all.
$m > 1$: $P = p \cdot fv(m)$
$m < -1$: $P = p \cdot [fv(-m) - 1]$
where $fv(m)$ is the value of the mth free variable.

At first, this may look strange or complicated, but a few examples will show that this concept is in fact quite straightforward, flexible and versatile:

The case for $m = 0$ is trivial and describes a refinable parameter P as possessing the starting value p. This is the normal case of the free refinement of a parameter.

If m is unity, the value of p is fixed. Assume you want to constrain an atom to lie on a mirror plane parallel to the $a - b$ plane at $c = -\frac{1}{4}$. The task is to fix the value for the z coordinate to -0.25. According to the above, this can be done by giving m the value of 1, and the value for p should be the atomic parameter of $z(-0.25)$. Hence, the atomic parameter for z in the .ins file for this atom reads 9.25.

To give a second example: Sometimes it can be helpful to fix the isotropic displacement parameter of an atom, U, at a certain value, for example 0.05. As always when parameters are fixed: $m = 1$; and p is the desired value for U: 0.05. The isotropic displacement parameter for the atom in question is then given as 10.05.

Whenever m is larger than 1 or smaller than -1, additional free variables are involved. As mentioned above, the most common case for this scenario is disorder. However CHIV and the distance restraints DFIX and DANG can also be combined with free variables. If, for example, the task is to restrain a ClO_4^- ion to be tetrahedral, this can be done with SADI restraints or with DFIX and an additional free variable.

Assuming the atoms in the ion are named Cl(1) and O(1) to O(4), the restraints using SADI are as follows:

```
SADI Cl1 O1 Cl1 O2 Cl1 O3 Cl1 O4
SADI O1 O2 O1 O3 O1 O4 O2 O3 O2 O4 O3 O4
```

The same can be achieved using DFIX and a second free variable. Applying the equations above, m is assumed to be 2 for the *second* free variable, and p is unity for the 1,2-distances and 1.633 for the 1,3-distances (taking into account that the 1,3-distances in a regular tetrahedron are 1.633 times as long as the 1,2-distances):

```
DFIX 21 Cl1 O1 Cl1 O2 Cl1 O3 Cl1 O4
DFIX 21.633 O1 O2 O1 O3 O1 O4 O2 O3 O2 O4 O3 O4
```

The value of the second free variable will be refined freely and converges at the mean Cl—O-distance. The advantage of the second way of restraining the ClO_4^- ion to be tetrahedral is that the average Cl—O distance will be calculated with a standard uncertainty (in addition to the individual Cl—O distances with their standard uncertainties). The disadvantage is that an additional least squares parameter has to be refined.

2.8 Results

As a result of a successful refinement, the crystallographer obtains a complete anisotropic model with all hydrogen atoms, which can be used to generate attractive figures for scientific publications (or grant proposals) and gain several kinds of information about a molecule. The most obvious are bond lengths and angles, but numerous other quantities can be calculated from the atomic coordinates, such as torsion angles or hydrogen bonds. The following paragraphs give a short overview about how to obtain which values (for a more detailed description consult the SHELXL manual). All values calculated by the program are generally accompanied by their estimated standard uncertainties as derived from the full correlation matrix.

2.8.1 Bond lengths and angles

If the command BOND appears in the header of the .ins file, SHELXL writes into the .lst file a table of all bond lengths and angles in the connectivity table. BOND $H expands this table to include all distances and angles involving hydrogen atoms as well.

2.8.2 Torsion angles

If the crystallographer or chemist wishes certain torsion angles to appear in a separate table in the .lst file, each of these torsion angles can be specified in a CONF command:

```
CONF atomnames
```

where atomnames defines a covalent chain of at least four atoms. If no atom names are specified, SHELXL generates all possible torsion angles.

2.8.3 *Atoms on common planes*

With the syntax

```
MPLA na atomnames
```

SHELXL calculates a least-squares plane through the first na atoms of the named atoms. The equation of this plane, together with the deviations of all named atoms from the plane and the angle to the previous least-squares plane (if present) are written into the .lst file. If na is not specified, the program fits the plane through all named atoms.

This command can also be used to determine the distance of an atom from a plane. If the task is to calculate the distance of a metal atom (say Ti(1)) from a Cp ligand (say atoms C(1)–C(5)), which coordinates to the metal via all five carbon atoms (typical $\eta 5$ style binding of Cp), the MPLA command looks like this:

```
MPLA 5 C1 C2 C3 C4 C5 Ti1
```

This calculates the best plane through the first five atoms (the Cp ligand) and the distance of all atoms mentioned (including the Ti) from that plane. All distances, together with standard uncertainties, are written into the .lst file.

2.8.4 *Hydrogen bonds*

If the command HTAB (without extensions) appears in the header of the .ins file, SHELXL performs a search over all polar hydrogen atoms[11] present in the structure and examines hydrogen bonding. The bonds listed in the .lst file are those for which the distance between acceptor and hydrogen atom are smaller than the radius of the acceptor atom plus 2.0 Å, and the angle between the donor atom, the hydrogen and the acceptor atom is larger than 110°.

With the syntax

```
HTAB donor-atom acceptor-atom
```

SHELXL generates hydrogen bonds with standard uncertainties and, in combination with ACTA (see below), the appropriate table in the .cif file. EQIV can be used to specify a symmetry equivalent of the acceptor atom. The third example in Chapter 3 deals with acidic hydrogen atoms and hydrogen bonds and the use of HTAB and EQIV is demonstrated there.

2.8.5 *The* **RTAB** *command*

The command RTAB allows the crystallographer to compile a variety of structural quantities. Depending on how many atoms are specified in the qualifier atomnames,

[11] That is hydrogen atoms bonded to electronegative elements.

```
RTAB codename atomnames
```

calculates and tabulates chiral volumes (one atom specified), distances (two atoms), angles (three atoms) or torsion angles (four atoms specified). `codename` must be specified and serves as an aid to identify the tabulated quantity in the .lst or .cif file. It must begin with a letter and cannot be longer than four characters.

2.8.6 *The* **MORE** *command*

The command `MORE m` sets the amount of output into the .lst file. `MORE 0` gives the least and `MORE 3` the most verbose output. The default value for `m` is 1.

2.8.7 *The .cif file*

The interface between the crystallographer and author of a scientific publication involving a crystal structure on one side and the reader of this publication as well as electronic databases on the other side is the 'Crystallographic Information File' (also known as the .cif file) as introduced by the International Union of Crystallography (Hall *et al.* 1991).

If the command `ACTA` appears in the header of an .ins file, SHELXL generates such a .cif file. `ACTA` automatically sets the `BOND`, `FMAP 2`, `PLAN` and `LIST 4` instructions and `ACTA` cannot be combined with other `FMAP` or `LIST` commands. Torsion angles defined by `CONF` and hydrogen bonds defined by `HTAB` are also written into the .cif file, while quantities defined by `RTAB` and `MPLA` are only tabulated in the .lst file.

2.9 Refinement problems

There are many more or less difficult problems a crystallographer can encounter when refining a crystal structure. The most prominent problems are twinning, disorder, pseudo-symmetry and atom type ambiguities. A whole set of additional difficulties is related to the refinement of protein structures.

The following chapters are intended to address the most common problems in a way that can easily be understood by scientists who have basic crystallographic knowledge and a minimum of experience refining simple crystal structures.

3

Hydrogen atoms

Hydrogen atoms are important in chemistry and are difficult to localize in an X-ray structure.[1] When passing through the crystal, X-ray photons interact with the electrons in the crystal, giving rise to the diffraction pattern, while the nuclei of the atoms do not contribute to the measured intensities. Hence it is electron density that we measure by X-ray diffraction. The heavier an atom is and the more electrons it has, the stronger its effect on the diffraction pattern. This also means that, especially in the presence of heavy atoms, light atoms are somewhat more difficult to localize. The lightest atom of all is hydrogen: it has only one electron, located away from the nucleus. Therefore, hydrogen atoms are notoriously difficult to detect with X-ray diffraction methods. Very accurate high quality data and proper scaling are required to distinguish hydrogen atoms from the background noise.

Especially for hydrogen atoms bound to carbon it is usually possible to calculate the hydrogen positions from the coordinates of the atoms to which the hydrogen atoms are attached, as the standard bond lengths and angles are well-known. Hydrogen atoms of water molecules, however, must be detected in the experimental electron density or else they may not be included into the model.[2] Even more difficult to detect can be hydrogen atoms in heavy metal hydrides. The sometimes relatively strong Fourier truncation ripples close to heavy atom positions can overpower the weak electron density maxima representing the hydrogen atoms. Very accurate and especially complete data, as well as a careful refinement, are required to localize those hydrogen atoms.

3.1 X—H bond lengths and U_{eq} values of H atoms

In X-ray diffraction, interatomic distances involving hydrogen atoms are always determined too short, for two reasons: first, the one electron belonging to the hydrogen atom is not observed at the actual site of the nucleus but, owing to the electronic interactions that actually make the bond, this electron is localized between the hydrogen atom and the atom to which the hydrogen is bound. This makes the bond appear shorter. The second effect is libration. As explained in detail in Chapter 8, thermal motion of atoms reduces the apparent bond distance among those atoms. This effect is stronger for light atoms, and particularly affects terminally bound atoms. Hydrogen

[1] With neutron diffraction methods, hydrogen atoms can be found very easily. However, neutron diffraction requires very large crystals and a neutron source.

[2] Sometimes the location of the hydrogen on water can be determined by inference from the surroundings (that means mostly finding partners for hydrogen bonds). This is particularly often the case for protein structures.

atoms are both very light and mostly terminally bound. Hence they are heavily affected by libration. Libration is temperature dependent (stronger motion at higher temperatures), which leads to longer observed X—H bond distances at lower temperature. At first, this observation seems to contradict common sense, which suggests that interatomic distances should be shorter at lower temperature. However this is not about true distances but about apparent ones, *observed* shorter due to libration, an effect much stronger at higher temperature. Therefore, when using standard X—H bond lengths to calculate hydrogen positions from the coordinates of other atoms, we have to take into account the temperature of the crystal during the diffraction experiment and assume slightly longer distances for lower temperatures (`TEMP` instruction in SHELXL).

As the location of hydrogen atoms is difficult to determine with accuracy, so is their thermal motion. We can, however, assume that the motion of a hydrogen atom is proportional to the motion of the heavier atom to which the hydrogen atom binds, and, as just elaborated upon, hydrogen atoms are affected by thermal motion even more strongly than heavier atoms. Furthermore, it is fair to assume that hydrogen atoms in methyl groups move somewhat more than other hydrogen atoms, as an additional degree of freedom, the torsion angle, is present. Therefore when including hydrogen atoms into our atomic model we generally do not refine their thermal motion, but assume that the isotropic U value of a hydrogen atom is 1.20 times the U_{eq} value of the atom to which the hydrogen binds. For methyl groups this factor is assumed to be 1.50. In SHELXL this is achieved by replacing the isotropic U value of the hydrogen atoms in the .ins file by -1.2 (or -1.5 for H atoms in methyl groups). The minus ties the U value to the U_{eq} of the non-hydrogen atom immediately preceding the hydrogen atom in the .ins file. Therefore it is crucial to correctly position the hydrogen atoms in the file, so that they directly follow the non-hydrogen atoms to which they bind.

3.2 Hydrogen bound to different atom types

The treatment of hydrogen atoms in an X-ray structure depends on many things, such as the geometry of the hydrogen atom containing moiety, the temperature of data collection, which influences the X—H distances, and the element type of the atom bound to hydrogen which also plays a role.

3.2.1 Hydrogen bound to carbon atoms

In most cases, the positioning of hydrogen atoms bound to carbon in an atomic model during the refinement of an X-ray crystal structure is done entirely without any or only very little direct information from the diffraction experiment. Sometimes, for example for the six hydrogen atoms of a benzene molecule, the location of the hydrogen atoms is obvious and can easily be calculated from the carbon positions. In other cases, for example for the three methyl hydrogen atoms of an ethyl group, the hydrogen positions are not quite obvious, but it can be expected that the torsion angle for the methyl group is such that the methyl H-atoms are staggered with respect to the other atoms in the ethyl group. This assumption allows us to calculate the hydrogen positions from the carbon positions alone, without any experimental information about the hydrogen atoms themselves. In other cases, such as the methyl groups in acetonitrile or toluene, the torsion angle of a methyl group cannot be calculated and it is impossible to find the hydrogen coordinates without experimental information. The problem of the torsion angle of a methyl group in acetonitrile, toluene or Cp*,[3] etc. however, is still relatively easy to solve: we know the C—H bond length and the H—C—H angle. That gives us a specific circle on which the three atoms must lie, and we even know the distances between the three atoms on that circle. We only need to determine the electron density function along that circle and find the slight maxima that correspond to the hydrogen atoms (Figure 3.1).

3.2.2 Hydrogen bound to nitrogen and oxygen

Even though standard bond lengths for N—H and O—H bonds are well-known and tabulated, it is frequently not clear whether a nitrogen or oxygen atom is protonated at all. In such cases the presence or absence of a hydrogen atom could change

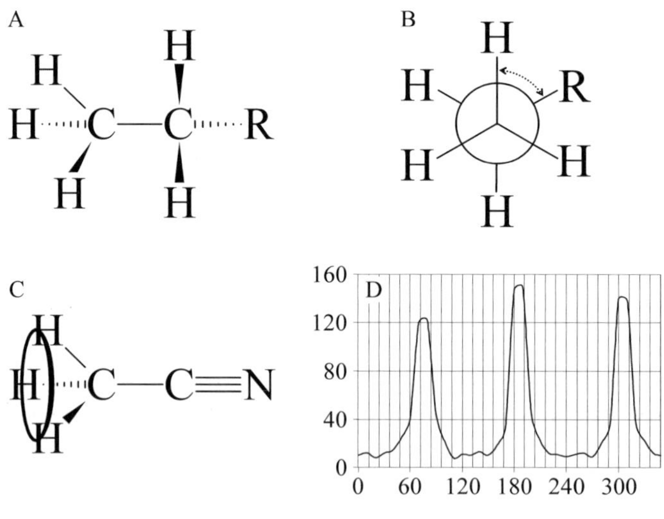

Fig. 3.1 Examples of hydrogen atom placement on methyl groups. **A**: the hydrogen atom positions in an ethyl group can be calculated from the carbon positions alone, when a staggered arrangement can be assumed. **B**: the same situation in the Newman projection, showing the torsion angle with a dotted-line arrow. **C**: Methyl group in acetonitrile: the circle through the hydrogen atoms corresponds to the line in space on which the hydrogen atoms must lie, regardless of the torsion angle. **D**: Electron density along this circle (simulated data on arbitrary scale; the horizontal axis gives the place on the circle in degrees from an arbitrary starting point): expected are three maxima about 120° apart, which correspond to the location of the hydrogen atoms on the circle in Figure 3.1.C.

[3] Pentamethyl-cyclopentadienyl

the determined oxidation state of a metal atom, which can be a crucial piece of information for the chemist. Therefore it is considered good practice to include potentially acidic hydrogen atoms into the model only when they can actually be found in the difference electron density. That means the crystallographer has to find significant residual electron density maxima on positions where a hydrogen atom could realistically be. If there is no such electron density, the hydrogen atom in question should not be included in the model. This does not mean that it is not present; it only means it cannot be found based on the diffraction data.

Sometimes, it is known that an oxygen or nitrogen must be protonated in order to achieve electro-neutrality; for example, when the oxidation state of all atoms and the charge of all ions in the structure are known or have been otherwise determined. In that case the hydrogen atoms on nitrogen and/or oxygen can be calculated as for carbon atoms. Hydroxyl groups pose the same theoretical problem described above for methyl groups: The exact position of the hydrogen atom cannot be calculated from the oxygen coordinates alone. However, the circle on which the hydrogen atom should lie can be predicted with some confidence. The accuracy of this circle is somewhat lower than for the hydrogen atoms in a CH_3 group, as the O—H bond distance can vary more than the C—H bond length.

3.2.3 Hydrogen bound to metals

Hydrogen atoms in metal hydrides are particularly difficult to localize in the difference electron density map, as Fourier truncation ripples from heavy atoms tend to obliterate the weak electron density of hydrogen atoms in their vicinity. Only very accurate and especially highly complete data to subatomic resolution (better than about 0.8 Å) is good enough to find hydrogen positions in the difference Fourier synthesis next to heavy metal positions. It may sound surprising that high-resolution data is required to localize hydrogen atoms, as, owing to their weak scattering and form factor, the contribution of the hydrogen atoms to data beyond about 1.5 Å is practically zero. However, the Fourier truncation ripples are much weaker and located closer to the metal atom with high-resolution data (see Cochran and Lipson 1966). This makes it much easier to distinguish the residual electron density maxima corresponding to hydrogen atoms from noise and spurious electron density.

3.3 Placing hydrogen atoms in SHELXL

In a real-life refinement, the X—H bond lengths and H—X—H angles are usually given as constraints. SHELXL makes the determination of hydrogen positions easy: in addition to generating the hydrogen atoms at the correct positions, the HFIX command generates all necessary constraints for any given C—H and most other X—H situations. The general syntax of the HFIX command is

```
HFIX mn atomnames
```

where m describes the geometry and defines the number of hydrogen atoms to be generated and n tells the program how to treat the hydrogen atom or atoms. The atoms that carry the hydrogen atoms are specified by `atomnames`. `HFIX` calculates the appropriate hydrogen positions, generates the hydrogen atoms and introduces the correct `AFIX` constraints necessary for the refinement with the hydrogen atoms in question (see Chapter 2 for a general description of the `AFIX` command). In the .res file, the newly introduced hydrogen atoms are preceded by a line `AFIX mn` and followed by a line `AFIX 0` to conclude the section of hydrogen atoms. The isotropic *U* values for the newly introduced hydrogen atoms are automatically replaced with −1.2 (−1.5 for methyl groups) by the program. SHELXL determines the appropriate X—H bond lengths for the temperature specified in the .ins file (`TEMP` followed by the temperature in degrees Celsius). Therefore it is important to specify the crystal temperature at which the data were collected.

In most cases n is going to be 3, which describes a 'riding model'. This model treats a hydrogen atom like a rider on a horse, where the horse is the non-hydrogen atom. When the non-hydrogen atom moves during the refinement, the hydrogen atom follows accordingly, as a person on a horse follows the horse when it moves about (assuming the person would not fall off the animal). Other values for n frequently used for hydrogen atom refinement are 7 and 8. Both also describe riding models, however with additional degrees of freedom (see below).

3.3.1 List of most common m *and* n *values in* HFIX *commands*

A general, clear and complete description of all possible values for the qualifiers m and n in `AFIX` constraints are given in the SHELX user manual. The following list mentions only the nine most common combinations of m and n values used for the generation of hydrogen atoms with the help of `HFIX`, and is not supposed to be complete.

`HFIX 13` Idealized *tertiary* C—H group with all X—C—H angles equal, subsequently refined using a riding model.

`HFIX 23` Idealized *secondary* CH_2 group with all X—C—H and Y—C—H angles equal, refined using a riding model. The H—C—H angle is calculated to be approximately tetrahedral, but is widened if X—C—Y is much less than tetrahedral.

`HFIX 33` Idealized CH_3 group with tetrahedral angles, refined using a riding model. The torsion angle of the methyl group is calculated to be staggered with respect to the shortest X—C bond. This requires the atom that carries the methyl group to be of tetrahedral geometry. If this is not the case (for example in a toluene or an acetonitrile molecule), `HFIX 33` cannot be used!

`HFIX 43` Aromatic C—H or amide N—H group, refined using a riding model. The hydrogen will be placed on the external bisector of the X—C—Y or X—N—Y angle.

`HFIX 93` Idealized *terminal* X=CH_2 or X=NH_2^+ group, refined using a riding model. The hydrogen atoms are placed to lie in the plane of the nearest substituent on X.

`HFIX 123` Idealized *disordered* CH_3 group. Like `HFIX 33`, but two alternative positions of the methyl group are calculated, rotated from one another by 60°. The resulting model will have 6 hydrogen atoms, each with 50% occupancy.

`HFIX 137` Idealized CH_3 group with tetrahedral angles. The initial torsion angle of the methyl group is determined via a difference Fourier analysis, and a rigid group refinement is performed for the methyl group, determining the best torsion angle, while retaining tetrahedral geometry. This is the most accurate and elegant way of calculating the hydrogen coordinates of a methyl group; however, it requires the data to be accurate enough to show at least slight maxima on the actual hydrogen positions.

`HFIX 147` Idealized OH group with tetrahedral X—O—H angle. As with `HFIX 137` the initial torsion angle is derived from a difference Fourier synthesis and a rigid group refinement is performed.

`HFIX 163` Acetylenic C—H with X—C—H linear, refined using a riding model.

3.3.2 Semi-free refinement of acidic hydrogen atoms

As mentioned above, it is not always easy to calculate the positions of hydrogen atoms bound to nitrogen and oxygen. However, when good low-temperature data are available, the hydrogen atoms can frequently be found in the difference Fourier synthesis and their coordinates can simply be taken from the list of residual density maxima at the end of the .res file. The so found hydrogen atoms can either be refined using `AFIX` constraints, or in a semi-free way by only *restraining* the X—H distances using `DFIX`. This restraint requires the crystallographer to specify a target value for the distance. As explained in Section 3.1, apparent X—H distances in X-ray structures are slightly longer at lower temperature, but altogether significantly shorter than the true distances between the nuclei. The .lst file contains a table with appropriate X—H bond distances at the temperature specified in the .ins file.[4] The target values for `DFIX` can be taken from there. The third example in this chapter deals with such a case.

This approach of semi-free treatment of acidic hydrogen atoms is elegant and allows for a somewhat less restricted refinement of the hydrogen position. In very rare cases, when extremely accurate high-resolution data is available and a hydrogen

[4] This table is only generated, when at least one hydrogen atom refined with the help of an `AFIX` constraint is present in the model.

atom is involved in a strong hydrogen bond, which significantly lengthens the donor-hydrogen distance, it may even be possible to refrain from using a distance restraint altogether. In any case, the isotropic U values for a semi-freely refined hydrogen atom should still be constraint to 1.2 times the U_{eq} value of the N or O atom the hydrogen binds to.[5]

3.4 Hydrogen bonds in SHELXL

SHELXL can analyze and tabulate hydrogen bonds. The bond lengths and angles of all non-hydrogen atoms are tabulated in the .lst file if the `BOND` command is present in the .ins file.[6] If `BOND $H` is present, bonds and angles involving all hydrogen atoms are also printed. Hydrogen bonds, however, are not automatically included into the .lst or .cif file but need to be specified using the `HTAB` or `RTAB` commands. More details about `BOND`, `HTAB` and `RTAB` can be found in Chapter 2. The third example in this chapter deals with acidic hydrogen atoms and the use of `HTAB`. Another case where `HTAB` can be used is the second example in Chapter 6.

3.5 Examples

In the following sections I present three examples of hydrogen atom placement and treatment. All files you may need in order to perform the refinements yourself are given on the CD-ROM that accompanies this book. In the first example the five most common `HFIX` commands (`HFIX 13`, `23`, `33`, `43`, and `137`) are used to place hydrogen atoms bound to carbon. The second example deals with the localization of hydrogen atoms in a metal hydride, while the third case is about acidic hydrogen atoms involved in hydrogen bonds. This last example also introduces the practical use of `HTAB` and `EQIV`.

3.5.1 Routine hydrogen atom placement: $C_{31}H_{54}MoN_2O_2$

$C_{31}H_{54}MoN_2O_2$ crystallizes in the monoclinic space group *Pn* with one molecule in the asymmetric unit (Adamchuk *et al.* 2006). The refinement of this structure was straightforward and by all means a routine case. There are no disorders or co-crystallized solvent, and the molecule possesses several different kinds of hydrogen atoms. All this makes this molecule a good example for routine hydrogen placement. The file hyd-01.res on the accompanying CD-ROM contains a complete anisotropic model without the hydrogen atoms. Figure 3.2 shows the complete molecule giving the name of all non-hydrogen atoms. Based on the figure, we can identify the different types of C—H species in the structure (the distance between

[5] It should be mentioned that not everybody in the crystallographic community agrees that the U value of hydrogen atoms must always be constrained relative to the U_{eq} value of the atom the hydrogen binds to. Sometimes, with very good data, one can possibly let the U value of the one or the other hydrogen atom refine freely.

[6] This is usually the case when the original .ins file for SHELXS has been generated automatically with a program like XPREP.

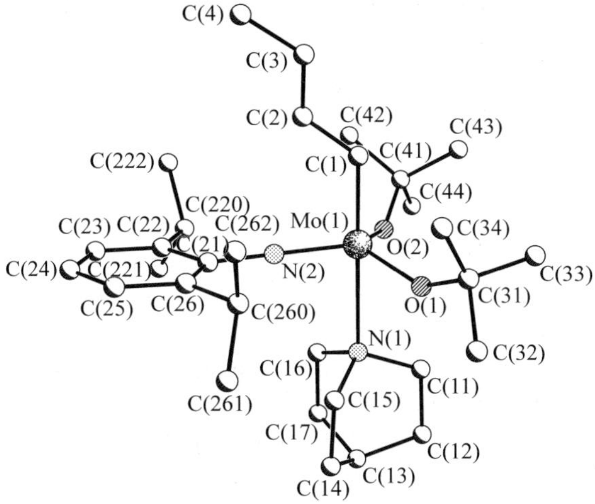

Fig. 3.2 Molecular structure of $C_{31}H_{54}MoN_2O_2$ with atomic labeling scheme.

C(2) and C(3) is 1.33 Å (a typical C—C double bond), and the bond between Mo(1) and C(1) should also be considered a double bond):

tert-CH	C(13), C(220), C(260)
sec-CH_2	C(11), C(12), C(14), C(15), C(16), C(17)
CH_3	C(4), C(32), C(33), C(34), C(42), C(43), C(44), C(221), C(222), C(261), C(262)
C(sp^2)-CH	C(1), C(2), C(3), C(23), C(24), C(25)

This covers all hydrogen atoms. According to the list in 3.3.1, we can choose the correct `HFIX` code to have SHELXL generate the hydrogen atoms and the appropriate `AFIX` constraints. The `HFIX` code for tertiary CH groups is `13`, for secondary CH_2 groups `23` and for aromatic hydrogen atoms `43` (for the purpose of hydrogen atom generation, the atoms C(1), C(2) and C(3) can be treated as aromatic carbons, as the geometrical situation is equivalent). As all but one methyl groups are bound to sp^3 carbon atoms, we have the choice of how to place the hydrogen atoms: either on purely calculated positions, staggered with respect to the shortest X—C bond (`HFIX 33`) or each CH_3 moiety as a rigid group, refining the torsion angles (`HFIX 137`). Which version is better depends mostly on the data quality: if the data are good enough to contain information about the actual hydrogen positions, it is a good idea to refine the torsion angles, if not, `HFIX 33` is more robust. As mentioned above, `HFIX 33` should only be used, if the methyl group in question is bound to a tetrahedrally coordinated atom, as other situations do not allow staggering the hydrogen atoms (see 3.2.1 and Figure 3.1), therefore we have no choice but to use `HFIX 137` with the hydrogen atoms on C(4). When you examine the residual electron density in the file hyd-01.res with a program like XP or Ortep, you will see that many of the residual density maxima correspond to hydrogen positions.[7] This makes it likely that `HFIX 137` will work, in which case, we should prefer it. Let's try the staggered version first, and second the somewhat freer version, and then compare. Edit the file hyd-01.res adding the following five lines directly before the first atom:

```
HFIX 13 C13 C220 C260
HFIX 23 C11 C12 C14 C15 C16 C17
HFIX 33 C32 C33 C34 C42 C43 C44 C221 C222 C261 C262
HFIX 43 C1 C2 C3 C23 C24 C25
HFIX 137 C4
```

Then save the file as hyd-02.ins and run SHELXL.

After some 10 cycles the R-values have dropped to $R1 = 0.0316$ (for $F > 4\sigma(F)$) and $wR2 = 0.0863$ (all data)[8] and our model now contains all hydrogen atoms, included at their calculated positions, refined using a riding model. Now, for a comparison, go back to the file hyd-02.ins, change the `HFIX 33` to `HFIX 137`, save as hyd-03.ins and re-run SHELXL. This time, after 10 cycles the R-values have dropped to $R1 = 0.0315$ (for $F > 4\sigma(F)$) and $wR2 = 0.0860$ (for all data). The only difference in the two refinements is that we refine the torsion angles of the 10 methyl groups bound to sp^3 carbons, and the R-values are slightly better. Of course this could be a merely calculative effect, as including more parameters into a refinement always improves the R-values. It is also possible that the two models are highly similar, as the torsions angles could very well have refined to values very close to the ones calculated by `HFIX 33`. After all it is relatively likely that the methyl groups are in fact staggered with respect to the neighbour atoms. Yet even if the two models should be indistinguishable, if the data are good enough to refine the torsion angles, it is a good idea to let them refine, as the resulting hydrogen positions reflect more accurately the calculated electron density maxima. To check whether there are actual differences in electron density for different torsion angles, we should look into the list file hyd-03.lst: towards the beginning of this file there is a list with an entry for every methyl group refined with `HFIX 137`. The first two entries look like that:

```
Difference electron density (eA^-3x100) at 15 degree intervals for
AFIX 137 group attached to C4
The center of the range is eclipsed (cis) to C2 and rotation is
clockwise looking down C3 to C4

112 91 54 34 28 34 49 57 63 63 46 26 26 33 49 68 63 42 21 10 18 42 70 98

Difference electron density (eA^-3x100) at 15 degree intervals for
AFIX 137 group attached to C221
The center of the range is eclipsed (cis) to C22 and rotation is
clockwise looking down C220 to C221

8 41 56 59 67 59 31 11 11 34 66 85 85 69 47 25 8 16 50 77 75 50 12 -7
```

[7] The location of Q(19) corroborates the interpretation of the carbon–molybdenum interaction as double bond and justifies the use of `HFIX 43` for C(1).

[8] The R-values have been defined in Equations 2.3 and 2.4.

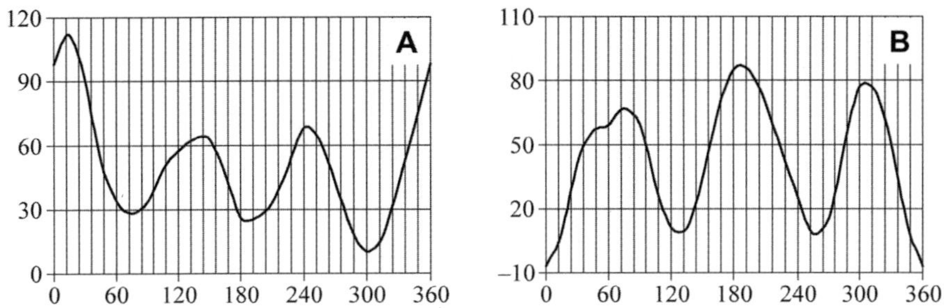

Fig. 3.3 Electron density along the circle through the three hydrogen atoms **A**: bound to carbon C(4) and **B**: bound to carbon C(221). In both cases three clear maxima are visible, which are about 120° apart.

The 24 numbers in each entry correspond to difference electron density along the circle of rotation of the methyl group, as described in Figure 3.1. We expect to see three maxima, about 120° apart. Using the numbers from the .lst file, we can generate diagrams such as the one in Figure 3.1D. The diagrams for the first two entries are shown in Figure 3.3. It is apparent, that the electron density along the circle on which the three hydrogen atoms of the two methyl groups must lie, shows three clear maxima, about 120° apart. This means that the torsion angle can be determined with a relatively high accuracy. Examination of the other corresponding entries in the .lst file tells us that all torsion angles can be determined with comparable precision, which means that we can trust the hydrogen positions in the model corresponding to hyd-03.res.

All that is left to do with this structure is to refine the weighting scheme to convergence, which has been done in hyd-04.res.

3.5.2 Hydrogen atoms in a Zr-hydride

The ansa-zirconocene dihydride $[\{HN(SiMe_2C_5H_4)_2Zr(\mu\text{-}H)H\}_2]$ crystallizes in the monoclinic space group $P2_1/m$ with half a molecule in the asymmetric unit; the other half is generated by the crystallographic mirror (Bai *et al.* 2000). The file zrh-01.res contains a complete anisotropic model, which includes all hydrogen atoms, except the hydrogen atoms bound to the metal atoms. The structure also contains a disordered toluene molecule, which may be an interesting subject to practice disorder refinement. The inclined reader may delete the atoms of the toluene and try to refine the disorder from scratch using the techniques described in Chapter 5.

The list of residual electron density maxima at the end of the file zrh-01.res contains several peaks, located close to one of the two Zr atoms. It is interesting to note that the two Zr atoms lie on the crystallographic mirror. That means bridging hydrogen atoms could also lie on this mirror, which would make it even more difficult to locate them, as the asymmetric unit would contain only half of each hydrogen atom. When you look at the structure with the Q-peaks using a program like XP or Ortep, you can see that of the 16 residual electron density maxima near the Zr atoms

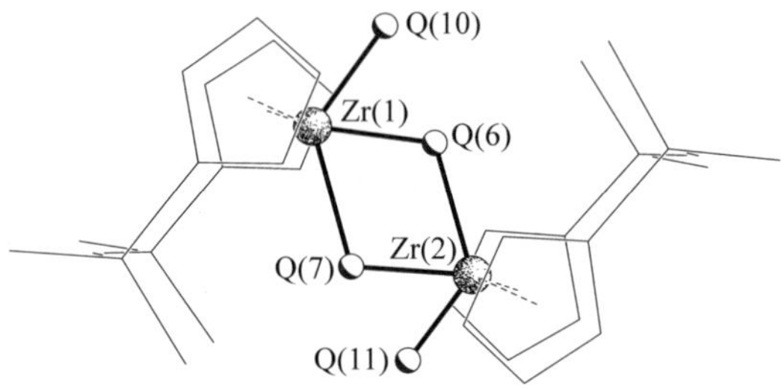

Fig. 3.4 Molecular structure of the zirconocene dihydride in a projection along the crystallographic *b* axis (that is looking onto the crystallographic mirror plane) with the four residual electron density maxima corresponding to hydrogen atoms.

(Q(1)–Q(12), Q(14), Q(15), Q(17), and Q(20)), eight are too close to the metal atoms (Q(1)–Q(4), Q(5), Q(8), Q(9), and Q(15))[9] and two (Q(12) and Q(17)) are too far away from the Zr centres to be considered as potential hydrogen atoms. Q(14) and Q(20) are relatively weak and not located at positions where we would expect hydrogen atoms to be. This leaves us with four candidates: Q(6), Q(7), Q(10), and Q(11). Q(6) and Q(7) are located precisely where we would have predicted bridging hydrogen atoms to be, and Q(10) and Q(11) could very well be terminal hydrogen atoms on the Zr centres. All four residual density maxima lie on the crystallographic mirror. Figure 3.4 shows the molecule with the four residual electron density maxima.

Assuming that the bridging hydrogen atoms are present, the presence or absence of the two terminal hydrogen atoms makes the difference between Zr(IV) and Zr(III). Zr(III) compounds are generally dark green to black, while most Zr(IV) containing molecules are of very light colour or entirely colourless. The crystals of this compound were pale yellow, which suggests Zr(IV) and the presence of both the bridging and the terminal hydrogen atoms. Therefore, turn the four Q peaks from the file zrh-01.res into hydrogen atoms (Q(6) ➔ H(1B), Q(7) ➔ H(2B), Q(10) ➔ H(1T), and Q(11) ➔ H(2T)) and place them next to the corresponding Zr atoms. Do not forget to change the values for U_{eq} to -1.2. The relevant section of the new .ins file looks like this:

```
ZR1  5  0.421613  0.250000  0.920000  10.5000   0.02136  0.01533=
       0.01863  0.00000  0.00188  0.00000
H1B  2  0.3893 0.2500 1.0450 10.5000  -1.2
H1T  2  0.6208 0.2500 1.0123 10.5000  -1.2
ZR2  5  0.127362  0.250000  1.062552  10.5000   0.01889  0.01493=
       0.02026  0.00000  0.00219  0.00000
H2B  2  0.1436 0.2500 0.9371 10.5000  -1.2
H2T  2 -0.0420 0.2500 0.9807 10.5000  -1.2
```

We do not know precisely to which target values to restrain the Zr—H distances, but we can at least make equivalent distances equivalent, using the SADI command:

[9] The atomic covalence radius of Zr is ca. 1.45 Å, that of hydrogen ca. 0.40 Å. This makes Q peaks 1.6 Å or less away from a Zr atom unlikely candidates for hydrogen atoms bound to Zr.

```
SADI ZR1 H1T ZR2 H2T
SADI ZR1 H1B ZR2 H2B
SADI ZR1 H2B ZR2 H1B
```

Save the file as your new .ins file. This corresponds to zrh-02.ins on the CD-ROM.[10]

After 10 cycles of least squares refinement, the model is complete and the weighting scheme can be adjusted and refined to convergence (that has been done in zrh-03.res).

3.5.3 Acidic hydrogen atoms and hydrogen bonds

The natural compound iromycine ($C_{19}H_{29}NO_3$) crystallizes in the tetragonal space group $I\bar{4}$ with one molecule of iromycine and one ethanol molecule in the asymmetric unit. We join the refinement at a point, where all non-hydrogen atoms are refined anisotropically and all hydrogen atoms bound to carbon have been placed. The file hbond-01.ins corresponds to this model. Figure 3.5 shows the structure with all hydrogen atoms determined so far.

The crystal diffracted relatively weakly and the data had to be truncated at 0.9 Å. As a result the data-to-parameter ratio is not very good, which is the reason why

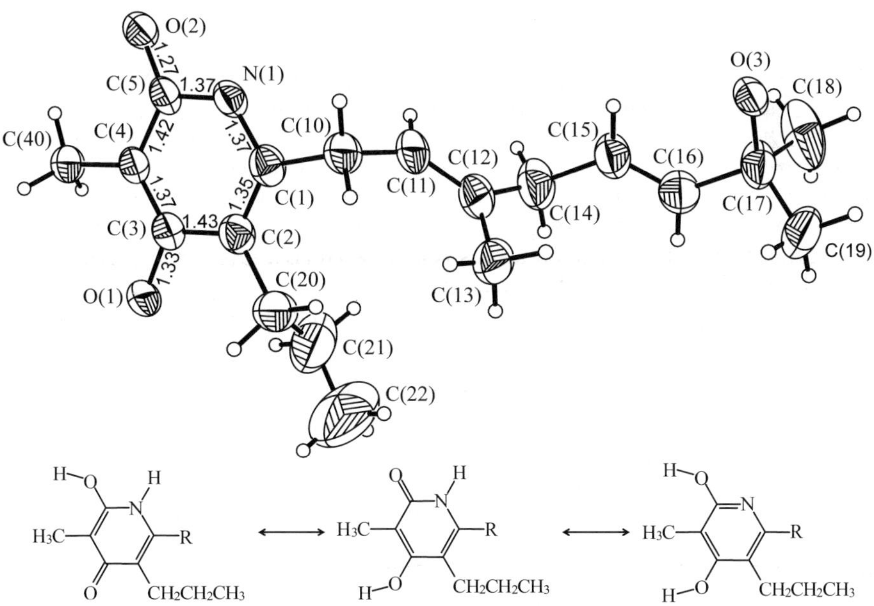

Fig. 3.5 Molecular structure of iromycine with all hydrogen atoms bound to carbon. The bond distances [Å] are given for part of the molecule. On the bottom of the figure are three of the possible tautomeric forms of the six-membered ring of the molecule.

[10] You may notice that in the actual file on the CD-ROM, the `SADI` commands are written in lower-case characters. As mentioned in Section 1.3.1, the SHELXL input is strictly not case-sensitive.

the ADP restraints `SIMU` and `DELU` have been applied to all atoms (see Chapter 2). There is one ethanol molecule near a special position, which has been refined using the `PART -1` instruction (see Chapter 5). The model as given in hbond-01.res and shown in Figure 3.5 is still missing the acidic hydrogen atoms. As explained above, it is not always easy to find those hydrogen atoms (especially with inferior data) and we rely on careful analysis of the residual electron density peaks combined with chemical knowledge. The latter tells us where to expect hydrogen atoms: oxygen atom O(3) should carry a hydrogen atom and two out of N(1), O(1), and O(2) should also be protonated, depending on which tautomeric form (as shown on the bottom of Figure 3.5) the molecule possesses. The bond lengths of the atoms involved in this tautomery (as tabulated in the file hbond-01.lst and also given in Figure 3.5) correspond more to one of the two keto-forms (middle in the figure), which makes us expect to find hydrogen atoms on O(1), O(3), and N(1). In addition, the oxygen atom in the ethanol molecule (not shown in Figure 3.5) should also be bound to a hydrogen atom.

Let's see what we can find in the difference Fourier: when you examine the file hbond-01.res with a program like Ortep or XP, you'll find that three of the 20 highest residual density maxima correspond to hydrogen atoms: Q(5) represents the hydrogen on N(1), Q(8) the hydrogen on O(3), and Q(14) the hydrogen on O(1). Q(14) is significantly lower than Q(5) and Q(8). This could mean that O(1) is only partially protonated, corresponding to a mixture of the two different keto-tautomers (middle and left on the bottom of Figure 3.5). Unfortunately our data does not appear to be good enough to resolve such a hydrogen-disorder, which is why we assume only one isomer, the one protonated on N(1), O(1), and O(3).

To make hydrogen atoms out of the three residual electron density maxima, do the following: copy each of the three Q-peaks in question from the file hbond-01.res *directly under* the atom the hydrogen atom binds to. Change the atom names (Q(5) ➔ H(1N), Q(8) ➔ H(3O), and Q(14) ➔ H(1O)) and the atom type from carbon to hydrogen. Also set the U_{eq} values of the three atoms to -1.2 as described in Section 3.1 of this chapter. To restrain the X—H bond lengths to sensible values (at this temperature: 0.84 Å for O—H and 0.88 Å for N—H), include two `DFIX` commands (one for the N—H distance, one for the two identical O—H distances). As explained in Section 3.3.2 the target values for the two `DFIX` restraints can be found in a table in the .lst file (search for 'default effective X—H distances'). Finally, rename the file to hbond-02.ins; the relevant sections of the new .ins file look like that:

```
DFIX 0.84 o1 h1o o3 h3o
DFIX 0.88 n1 h1n

WGHT     0.100000
FVAR        0.13214
O1    4    1.240355   0.247935   0.155510   11.00000   0.03230   0.05124=
         0.06363    0.01067    0.00258   -0.00420
H1O   2    1.2991 0.2192 0.1551 11.00000   -1.2
O2    4    1.129906 -0.003601   0.285825   11.00000   0.03637   0.04137=
         0.07358    0.00872   -0.00565   -0.00278
O3    4    0.428819   0.227755   0.142812   11.00000   0.03103   0.05653=
```

```
         0.04234    0.00364    0.00230   -0.00414
H3O  2    0.4562 0.2693 0.1705 11.00000   -1.2
N1   3    1.032006   0.116656   0.253000   11.00000   0.02779 0.04145=
         0.05920    -0.00131    -0.00093     0.00170
H3O  2    0.9879 0.0823 0.2716 11.00000   -1.2
```

After 10 refinement cycles, the three new hydrogen atoms are now part of the model and the *R*-values have dropped. We are, however, still looking for the hydrogen atom on the disordered ethanol molecule. As the asymmetric unit contains only half a solvent molecule, we are chasing half a hydrogen atom, corresponding to half an electron, which is problematic, to say the least. In addition the ethanol molecule seems to be subjected to relatively strong thermal motion, as the large thermal ellipsoids of the atoms of the solvent molecule tell us. That means that the missing half electron is probably not very well localized.[11] A close examination of the residual electron density in Ortep or XP reveals that Q(16) could be the hydrogen atom bound to O(1S). This is by no means certain, but having no real alternative, we will go for it, at least for the moment being.

For the file hbond-03.ins: copy Q(16) directly under oxygen O(1S), rename it to H(1OS), change atom type number and U_{eq} and expand the `DFIX 0.84` command to accommodate the newly included atom. In addition include the command `HTAB` into the file, somewhere after the `UNIT` card and before the first atom. As explained above, this will examine all hydrogen atoms bound to electronegative atoms for hydrogen bonding.

After 10 refinement cycles, the *R*-values have dropped again slightly and the file hbond-03.lst contains the following table at the end of the file, directly after the bond lengths and angles, assessing the hydrogen bonding patterns:

```
Hydrogen bonds with H..A < r(A) + 2.000 Angstroms and <DHA > 110 deg.

D-H       d(D-H)  d(H..A)  <DHA    d(D..A)  A
O1-H1O    0.841   1.882    145.76   2.620   O3   [x+1, y, z]
O3-H3O    0.848   1.862    165.87   2.693   O2   [y+1/2, -x+3/2, -z+1/2]
N1-H1N    0.854   2.022    154.51   2.818   O2   [-x+2, -y, z]
O1S-H1OS  0.871   1.998    116.64   2.512   O1S  [y+1, -x+1, -z]
```

This table specifies the names of the donor and hydrogen atoms in the first column, gives the donor–hydrogen and hydrogen–acceptor distances in the second and third column, the donor–hydrogen–acceptor angle in the fourth, the donor–acceptor distance in the fifth, and the name of the acceptor atom in the sixth column. In the last column a symmetry operator is given, if the hydrogen bond is to an acceptor atom in a different asymmetric unit. To analyze the first entry of the table in detail: the hydrogen atom H(1O), bound to O(1) has a distance of 0.841 Å to O(1) and a distance of 1.882 Å to the acceptor atom. The O(1)–H(1O)–acceptor angle is 145.77°, the O(1)–acceptor distance is 2.620 Å. The acceptor atom is the symmetry equivalent

[11] It also tells us that the hydrogen atom on the ethanol is probably involved in a hydrogen bond with a symmetry equivalent of the same solvent molecule, as a hydrogen bond to a well-behaved atom would restrict the motion of the ethanol.

of O(3), which is generated by the symmetry operator [x+1, y, z]. The other lines in the table are to be read similarly.

You may have noticed that all quantities in the table are given without standard uncertainties. As mentioned in Chapter 2, there is a second way of using the HTAB command: specifying the donor and acceptor atom together with HATB, generates all standard uncertainties. This is very easy for hydrogen bonds between atoms in the same unit cell. For hydrogen bonds to symmetry equivalent atoms, the symmetry operators need to be specified with the help of the EQIV command. In our case, we have four different hydrogen bonds involving four different symmetry operators. This gives rise to the following four EQIV and HTAB commands:

```
EQIV $1 x+1, y, z
EQIV $2 y+1/2, -x+3/2, -z+1/2

EQIV $3 -x+2, -y, z
EQIV $4 y+1, -x+1, -z

HTAB O1 O3_$1
HTAB O3 O2_$2
HTAB N1 O2_$3
HTAB O1S O1S_$4
```

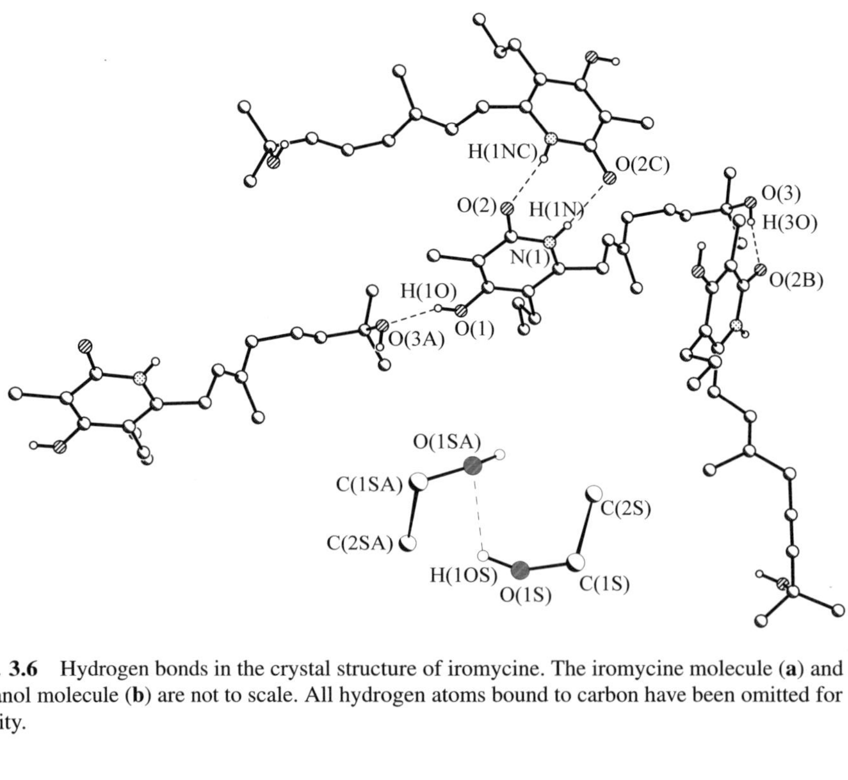

Fig. 3.6 Hydrogen bonds in the crystal structure of iromycine. The iromycine molecule (**a**) and the ethanol molecule (**b**) are not to scale. All hydrogen atoms bound to carbon have been omitted for clarity.

Include these eight lines into the new .ins file (hbond-04.ins) and run a refinement. The resulting list file, hbond-04.lst, contains a new table, directly next to the previous one:

```
Specified hydrogen bonds (with esds except fixed and riding H)

 D-H         H...A       D...A       <(DHA)
 0.84(2)     1.88(4)     2.620(4)    146(6)     O1-H1O...O3_$1
 0.85(2)     1.86(2)     2.693(5)    166(5)     O3-H3O...O2_$2
 0.85(2)     2.02(3)     2.818(5)    155(5)     N1-H1N...O2_$3
 0.87(2)     1.99(12)    2.51(3)     117(11)    O1S-H1OS...O1S_$4
```

This is all the information about the hydrogen bonds you can get from a standard crystal structure. The only thing left to do is the refinement of the weighting scheme to convergence. This has been done in the file hbond-05.res. Figure 3.6 shows the final model with some symmetry equivalent molecules and all hydrogen bonds.

4

Atom type assignment

4.1 All electrons are blue

Research chemists explore new reactions and sometimes entirely new types of molecules. A crystal of a new substance may have an unexpected structure, or an entirely unintended composition. After solving a structure, the crystallographer has to find a sensible interpretation of this solution, as the solution of the phase problem is a set of phases, resulting in an electron density map of unidentified peaks. Only in rare cases, all electron density maxima determined by the program used to solve the structure—for chemical structures almost always SHELXS—are assigned the correct atom types and the crystallographer needs to find out which density maximum corresponds to which chemical element. Therefore, in addition to an understanding of the diffraction experiment, the crystallographer needs a well-founded chemical knowledge. In the course of the diffraction experiment the X-ray photons passing through the crystal interact with the electrons of the atoms in the crystal, giving rise to the diffraction pattern, which consists of reflections. Each reflection corresponds to a structure factor $F(hkl)$, which can be understood as a Fourier summation over all atoms in the unit cell:

$$F(hkl) = \sum f \exp[2\pi i(hx + ky + lz)] \tag{4.1}$$

Reversing the Fourier transformation leads back to the electron density ρ.

$$\rho(x, y, z) = \frac{1}{V} \sum_h \sum_k \sum_l F(hkl) \exp[-2\pi i(hx + ky + lz)].^1 \tag{4.2}$$

The value of ρ at a given point (x, y, z) in the unit cell depends directly upon the measured intensities $I \propto F^2$. The intensities, in turn, depend upon the relative position of all atoms in the unit cell, but also upon the exposure time, the intensity of the primary X-ray beam, the size and shape of the crystal, etc., thus making the results of the experiment accurate only in a relative and not an absolute way. This means that without proper scaling the electron density as determined by the X-ray diffraction experiment is very hard to interpret. In order to assign meaningful dimensions like *electrons per cubic Ångström* to the density values, the measured intensities (or F^2 values in the .hkl file) have to be scaled properly.[2] This scaling

[1] It may strike the reader that there is no phase appearing in these equations. The phase Φ is 'hidden' in the atomic coordinates x, y, and z: $\Phi_i = 2\pi(hx_i + ky_i + lz_i)$.

[2] This is the purpose of the first free variable in SHELXL, which is also known as the *overall scale factor*.

depends largely on the model and hence the interpretation of the electron density itself, leaving us with a Catch 22. In many cases, the assignment of the atom type is easy and straightforward, for example, with phenyl rings or Cp*[3] ligands, where the geometry of the ligand gives away the nature of the atoms. In other cases, however, the atom type assignment can be extraordinarily difficult. Especially the distinction between elements that are relatively light and direct neighbours in the periodic table can sometimes be very hard,[4] but in some cases it can even be difficult to determine the nature of the only heavy atom.

The task of atom type assignment would be easy if the electron density obtained from the experimental intensities and initial phases were somehow colour-coded and electron density for carbon was, say, black, for nitrogen blue, for sulfur yellow, for oxygen red and so forth, but unfortunately all electrons are blue, at least on the computer screen.[5]

4.2 Chemical knowledge

A crystal structure must be chemically sensible or it is wrong. The number and lengths of the bonds as well as the coordination geometry of an atom in a structure, as well as the colour of the crystal, should be critically examined and taken into account. A carbon atom, for example, appearing to make five single bonds, is either part of a disorder or it is not a carbon atom, and if a crystal supposedly containing Cu(II) is colourless, other metals or Cu(I) should be considered. A sulfur–oxygen bond of 1.55 Å is too long to be credible (P–O would be more likely) and Pt(II) is very likely to be coordinated by four ligands in a square plane and not in a tetrahedron. There are many more examples of this kind and a good practical crystallographer needs a solid basis of chemical knowledge, or at least a periodic table and lists with bond lengths, preferred oxidation states and coordination geometries next to the diffractometer (see the tables in Chapter 12).

4.3 Crystallographic knowledge

The single most powerful crystallographic indicator for a wrongly chosen atom type is the refined U_{eq} value for this atom, which mirrors the size of the thermal ellipsoid (a sphere in the isotropic case). The thermal ellipsoid describes the space taken by a certain percentage—usually 50%—of the electrons of the refined atom. That means that in the case of, for example, an oxygen atom in a given model, a sphere is drawn around the volume increment of the electron density map corresponding to four electrons (50% of the eight electrons oxygen possesses). If this atom should indeed be oxygen, the size of the sphere will be comparable with the other spheres in the same structure. If, however, this particular atom was assigned incorrectly and is in

[3] Cp* is pentamethyl-cyclopentadienyl.

[4] Probably the most prominent example is the confusion of nitrogen and oxygen, which can show similar coordination geometries and which differ from one another only by one electron.

[5] If you are using XP to display your structures, you will find that electrons are actually green and not blue.

fact nitrogen, the sphere surrounding four electrons will be significantly larger. This is because nitrogen has one electron fewer than oxygen, and the volume increment corresponding to four electrons will be larger if the overall number of electrons is lower.

If, to give a second example, a tetrahydrofuran molecule in a structure is not coordinated to a metal atom, it can be difficult to tell which of the five positions corresponds to the oxygen. Refining all five positions as carbon atoms will result in four atoms that have about similar U_{eq} values and one atom with a significantly lower value for U_{eq}.[6] The atom with the smaller sphere is the oxygen atom, while the four about equal atoms are carbon. Figure 4.1 shows such a case.

In general, too small thermal parameters mean that the current model does not contain enough electrons at the place of an atom, while too large displacement parameters suggest the presence of a lighter atom than the one in the model.

If an atom is refined anisotropically, the displacement parameters will not be spherical but assume the shape of ellipsoids, corresponding to the thermal motion of the atoms. If the thermal ellipsoid of an atom is elongated, it can be assumed that this atom moves more strongly in one direction than in others. If the ellipsoid is strongly elongated (as in Figures 5.1 or 5.18), it is likely that this atom is involved in a disorder. Generally, the overall size of the thermal ellipsoids is as good a source of information about the element type as the isotropic displacement parameter. Very anisotropic-looking ellipsoids, however, can make it difficult to extract the size-information accurately. In such a case, the U_{eq} values calculated from the U^{ij} matrix elements can be used for comparison instead.[7] SHELXL writes the U_{eq} value for each atom into a table in the .lst file.

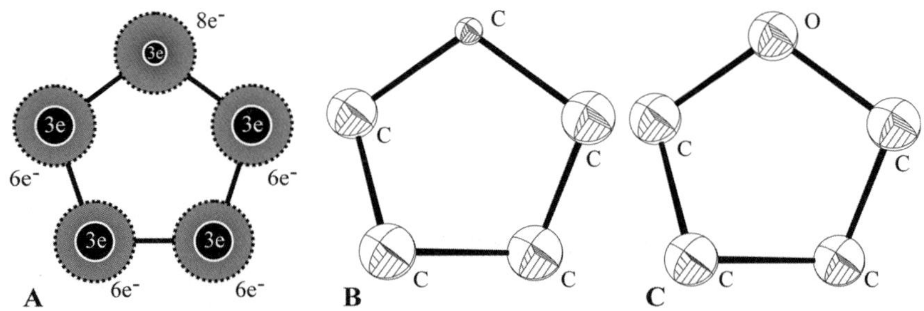

Fig. 4.1 **A**: Cartoon of a tetrahydrofuran molecule. The dashed circles in grey represent the atoms, the white rimmed black circles the volume increment corresponding to three electrons. **B**: Isotropic displacement parameter (50% representation) of a tetrahydrofuran molecule, where all atoms were refined as carbon; the sphere representing the true oxygen is much smaller than the other four spheres. **C**: The same molecule with correct atom type assignment; all five spheres are approximately equal in size.

[6] This is assuming that the thf is ordered. Sometimes, when the oxygen is disordered randomly over the five possible positions, it can be best to refine the tetrahydrofuran as a five-membered all-carbon ring. See Section 7.8.4 for an example.

[7] In the anisotropic case, U_{eq} is defined as a third of the trace of the orthogonalized matrix U^{ij}, which describes the anisotropic displacement-ellipsoid. Hence, U_{eq} mirrors the size of the thermal ellipsoid. More on anisotropic

4.4 Examples

In the following sections I present examples of cases in which the assignment of the atom type is not quite obvious. All files you may need in order to perform the refinements yourself are given on the CD-ROM that accompanies this book. The first example is a case of N–O distinction, where bond lengths and anisotropic displacement parameters as well as some chemical imagination help to find the right answer. The second case is somewhat less obvious and deals with the distinction between phosphorus and sulfur. The third example describes a structure where it was not possible to determine the nature of the only heavy metal in the molecule. This example makes clear how important proper scaling is for a well-determined structure.

4.4.1 Tetrameric $InCl_3$—the N or O question

Attempting to create a new InN precursor for CVD,[8] a chemist tried to eliminate Me_3SiCl from a mixture of Et_2NSiMe_3 and $InCl_3$ in tetrahydrofuran. Crystals suitable for a diffraction experiment were obtained from diethyl ether. X-ray structure analysis showed the crystals to consist of tetrameric $InCl_3$ in the triclinic space group $P\bar{1}$ with half a molecule per asymmetric unit. The rest of the molecule is generated by an inversion centre. The core of the structure consists of four In atoms linked by six bridging Cl atoms forming three four-membered In_2Cl_2 rings. In addition, the inner In atoms are each connected to one, and the outer In atoms to two terminal-bonded Cl atoms. To complete their distorted octahedral coordination sphere, the inner In atoms are coordinated by one and the outer ones by two solvent molecules, as shown in Figure 4.2. More details about this structure can be found in Müller *et al.* (2000).

It seemed obvious to refine the solvent molecules as diethyl ether, as the crystals had been obtained from this solvent. The refinement goes well and gives rise to good *R*-values. It is at this state that we join the refinement; the file in-01.res on the accompanying CD-ROM contains the complete anisotropic model with all hydrogen atoms on the ether molecules.

When you look at the atoms and difference density peaks in this file with a graphical interface such as XP or Ortep, the following becomes visible: the anisotropic displacement parameters of the oxygen atoms are slightly large when compared with the other atoms (average U_{eq} for O is 0.031 Å^2, and for the terminal Cl-atoms 0.026 Å^2). In addition, the C–O bond lengths measure on average 1.50 Å, which is longer than the expected standard value for carbon–oxygen single bonds (1.43 Å).

These two points could be explained in the following way: the oxygen atoms are part of solvent molecules, which are bound not very tightly to the In atoms. Thus somewhat larger anisotropic displacement ellipsoids can be expected, as the diethyl ether molecules are relatively free to move—in any case more so than

displacement parameters and nomenclature (e.g. why the eq in U_{eq} is subscript, while the *ij* in U^{ij} is superscript) can be found in Trueblood *et al.* (1996) and references therein.

[8] Chemical Vapour Deposition.

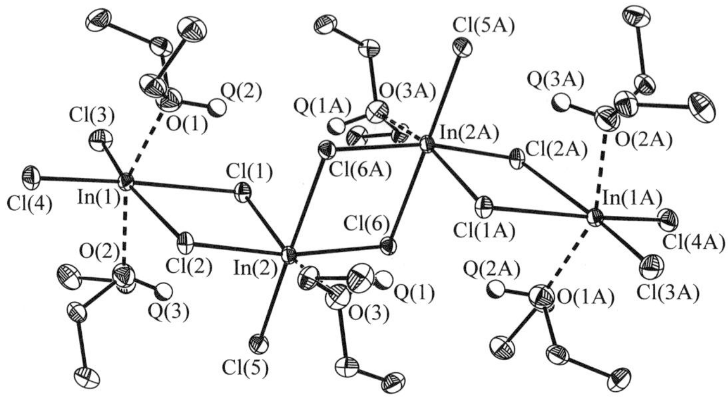

Fig. 4.2 Early model of tetrameric $InCl_3$ with Et_2O as coordinated solvent. Also shown are the three highest electron density maxima (Q(1), Q(2) and Q(3)) and their symmetry equivalents (there is a crystallographic inversion centre in the middle of the four membered ring formed by the atoms In(2), Cl(6), In(2A), and Cl(6)A). Hydrogen atoms have been omitted for clarity.

the terminal Cl atoms. The long oxygen–carbon distance could be explained with the coordination of the oxygen to the indium: following the bond-number concept of Linus Pauling (1947), the valence sum of all interactions for a given atom is constant, which makes existing bonds become longer when an atom undergoes new interactions.

A possible problem in this chain of argument is that the two explanations are mutually exclusive: either there is an In–O interaction that is sufficiently strong to elongate the C–O bond by almost 0.1 Å, or the molecule is not attached to the In and hence free to move enough to give rise to relatively large ADPs. Of course, both effects are not very strong and, if necessary, any attacks by referees can always be countered with the all-powerful argument of 'crystal packing'. Indeed, if the two aforementioned points (slightly larger oxygen ellipsoids and C–O bond distances) were the only peculiarities and no other corroboration for an incorrectly assigned atom type could be found, the diethyl ether version of this structure could pass as acceptable.

Yet there *is* more: even though the difference Fourier does not seem to be suspicious at first sight (highest maximum: 0.88 $eÅ^{-3}$, deepest hole: -1.36 $eÅ^{-3}$), the location of the three highest residual electron density maxima is striking: Q(1), Q(2), and Q(3) are localized close to the three independent oxygen atoms (O(3), O(1), and O(2), in this order), just as if they were hydrogen atoms (Figure 4.2). Spurious electron density can be found close to special positions and/or near heavy atoms as a result of Fourier chain truncation or absorption (more about this in Chapter 8), but none of the three highest residual electron density maxima is a candidate for this explanation.

Replacing diethyl ether with diethylamine in the model solves all the problems. Even though Et_2NH was used neither as a starting material in the original experiment nor during crystallization, traces of water could have found their way into the flask (tetrahydrofuran is known to be difficult to keep dry), hydrolyzing the Et_2NSiMe_3 to Et_2NH and the siloxane. The actual 'reaction' leading to the crystals was then:

$$6\ Et_2NH + 4\ InCl_3 \rightarrow [(InCl_3)_4 \cdot (Et_2NH)_6]$$

To implement this into the new .ins file, do the following: open the file in-01.res with a text editor and change the names of the three oxygen atoms to nitrogen by replacing the 'O' with an 'N' and—very important—by changing the atom type in the scattering factor list (SFAC) also from 'O' to 'N'. While you are at it, adjust the UNIT card to accommodate the new hydrogen atoms (three independent atoms in $P\bar{1}$ equals six atoms per cell).

```
SFAC C H O CL IN
UNIT 24 60 6 12 4
(...)
O1     3    -0.267483     0.310760     0.233650     11.00000
```

Is changed to

```
SFAC C H N CL IN
UNIT 24 66 6 12 4
(...)
N1     3    -0.267483     0.310760     0.233650     11.00000
```

Then make hydrogen atoms out of the three highest residual electron density maxima as explained in Chapter 3. Do not forget to include DFIX restraints for the three N–H distances, as it is needed for a semi-free refinement of acidic hydrogen atoms (see Section 3.3.2). Data were collected at −140°C, therefore the target value for the N–H distance should be 0.91 Å.[9]

All these changes have been made in the file in-02.ins, and a refinement with SHELXL gives rise to a much better model (in-02.res). In the final model (in-03.res; weighting scheme adjusted), the mean U_{eq} value for the nitrogen atoms is 0.017 Å^2, identical to the U_{eq} values of the bridging chlorine atoms in the same model. The average C–N bond length is still 1.50 Å and only slightly longer than a standard C–N single bond (1.47 Å). This elongation can easily be explained with the coordination of the diethylamine molecules to the In atoms. In addition all figures of merit improve significantly, and the fact that five crystallographically independent N–H–Cl hydrogen bonds can be found further corroborates the model. Figure 4.3 shows the final molecule with these hydrogen bonds.

When you examine the file in-03.ins carefully, you will find the following lines in the header:

```
htab

eqiv $1 -x, -y+1, -z
```

[9] See Section 3.3.2 and Example 3.5.3 for how and where to find the target value for the distance restraint.

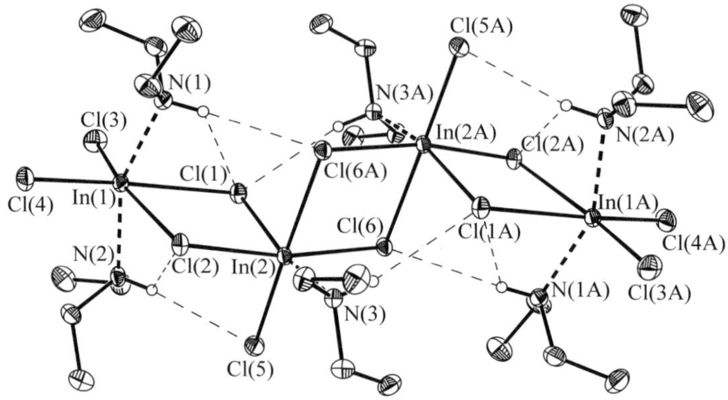

Fig. 4.3 Final model of the tetrameric $InCl_3$, with coordinating Et_2NH and N–H–Cl hydrogen bonds in the same orientation as in Figure 4.2. All hydrogen atoms bound to carbon have been omitted for clarity; the hydrogen atoms bound to nitrogen are at the positions of the residual density maxima shown in Figure 4.2.

```
htab n1 cl1
htab n2 cl5
htab n2 cl2
htab n1 cl6_$1
htab n3 cl1_$1

mpla 4 in1 cl1 in2 cl2
mpla 4 in2 cl6 in2_$1 cl6_$1
```

These lines have to be typed manually and generate informative output about the hydrogen bonds (`HTAB`) and the planarity of the four membered rings (`MPLA`) in the file in-03.lst. You can find this information towards the end of the .lst file, directly after the table with bond lengths and angles. The `EQIV` command is explained in detail in Chapter 3, `HTAB` is described in Chapters 2 and 3, and `MPLA` is explained in Chapter 2.

4.4.2 A cobalt salt

Trying to crystallize the Fe–S cluster protein FhuF, a protein excreted by *E. coli* bacteria grown under iron deficiency, a sparse matrix crystallization screening (Jancarick and Kim 1991) was performed in which the protein solution was brought into contact with many different agents, hoping that at least one of them would cause the protein to crystallize. Micro crystals were obtained from 1.8 M $(NH_4)_2SO_4$/0.01 M $CoCl_2$ at pH 6.5. These crystals could be grown to a size large enough for a diffraction experiment by means of repeated macro-seeding. The crystals turned out to contain no protein at all but to be of a cobalt salt. It crystallizes in the orthorhombic space group $Pmn2_1$ with half a formula unit per asymmetric unit, the rest is generated by the crystallographic mirror.

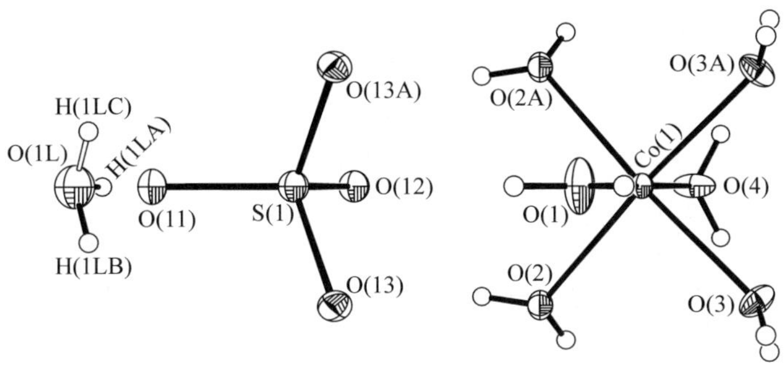

Fig. 4.4 Model of $[Co(H_2O)_6]SO_4{\cdot}H_2O$. The crystallographic mirror plane coincides with the atoms O(1L), H(1LA), O(11), S(1), O(12), O(1) (and the two hydrogen atoms bound to O(1)), Co(1), and O(4). The hydrogen atoms H(1LB) and its symmetry equivalent H(1LC) are refined with 50% occupancy, which is equivalent to a disorder between these two positions.

The identification of the Co(II) ion, which is octahedrally surrounded by six water molecules, was straightforward and easy. The tetrahedral counter ion was identified as sulfate, as the crystals had been grown in the presence of high amounts of ammonium sulfate. In addition to the two ions, one free water molecule was found in the difference Fourier synthesis, which, owing to its position on a crystallographic mirror, shows a peculiar positional disorder of its hydrogen atoms. This model is shown in Figure 4.4. The refinement went well and gave rise to excellent *R*-values. It is at this state that we join the refinement; the file co-01.res on the accompanying CD-ROM contains a complete anisotropic model of $[Co(H_2O)_6]SO_4{\cdot}H_2O$ with all hydrogen atoms on the water molecules.

Even though the values for the residual factors are exceptionally good ($R1 = 0.0189$ for $F > 4\sigma(F)$, $wR2 = 0.0518$ for all data),[10] there are two striking details in this structure, which need to be investigated: first the S–O bond distances (between 1.533(2) and 1.541(2) Å) are unusually long for sulfate, however just right for phosphate. Secondly it is unusual that disordered hydrogen atoms be so clearly visible in the difference electron density as found for the seventh water molecule.

The sulfate ion is probably phosphate and, unless we have an H_3O^+ ion in the structure, the seventh water should be an ammonium ion (the residual electron density maximum Q(3) in the file co-01.res conveniently lies very close to the possible position of the fourth hydrogen atom). Replacing the sulfur atom with phosphorus and changing the oxygen of the seventh water to nitrogen leads to the file co-02.ins (do not forget to adjust the `SFAC` and `UNIT` cards as well as the `DFIX` command for the N–H distances). Refinement with SHELXL gives rise to a new model with significantly improved figures of merit. The highest residual electron density maximum in co-02.res corresponds to the missing hydrogen atom on the ammonium ion. Including this atom into the model leads to co-03.ins (do not forget to change the `UNIT` card and the `DFIX` command). The ammonium ion in co-03.res has three independent

[10] The *R*-values have been defined in Equations 2.3 and 2.4.

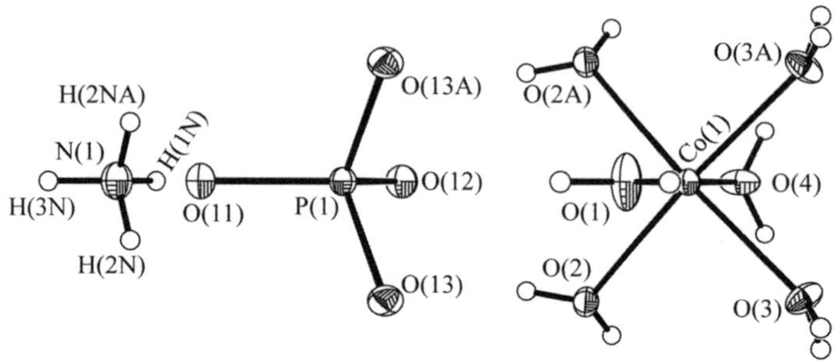

Fig. 4.5 Model of $[Co(H_2O)_6]NH_4PO_4$ in the same orientation as the molecule in Figure 4.4.

hydrogen atom positions, two of which lie on the crystallographic mirror. The file co-04.res contains the final model (shown in Figure 4.5); the weighting scheme has been adjusted and the final R-values are even lower than before ($R1 = 0.0152$ for $F > 4\sigma(F)$, $wR2 = 0.0399$ for all data).

The presence of an ammonium ion in the crystal does not come as a surprise, as the crystals were grown from a solution containing 1.8 M $(NH_4)_2SO_4$. A phosphate ion, however, would not have been expected in this structure. No phosphate-containing solution was used at any time during the purification, concentration or crystallization of the protein. On the other hand, phosphate is ubiquitous in nature and $CoNH_4PO_4$ is virtually insoluble. Therefore, in the presence of ammonium and cobalt, even traces of phosphate will lead to precipitation. The two seeding steps that were required to grow sufficiently large crystals could have introduced enough trace-phosphate from the protein solution to give the observed compound.

4.4.3 Unclear central metal atom

Using a spatula, a crystal was removed from a flask containing a silicon–nitrogen compound and mounted onto the diffractometer. This crystal was the only one to be found in the flask. The space group was determined to be $P\bar{1}$ with half a molecule per asymmetric unit. The model from SHELXS shows a silicon atom and several other electron density maxima. Ten of those maxima form a familiar shape: a Cp* ligand, and can be assigned the atom type carbon. The model in file si-01.ins on the accompanying CD-ROM reflects this interpretation of the solution.

After some 20 least squares cycles in SHELXL the model has not changed its geometry and the residual electron density maxima are rather low (highest peak 1.99 $e^-/Å^3$). However, the R-values are not good at all for a supposedly almost complete model, and the isotropic displacement parameters for the carbon atoms are very large (average U_{eq} is 0.264!), while the one for the silicon atom is relatively small (0.011). When you examine the list of residual electron density maxima, you

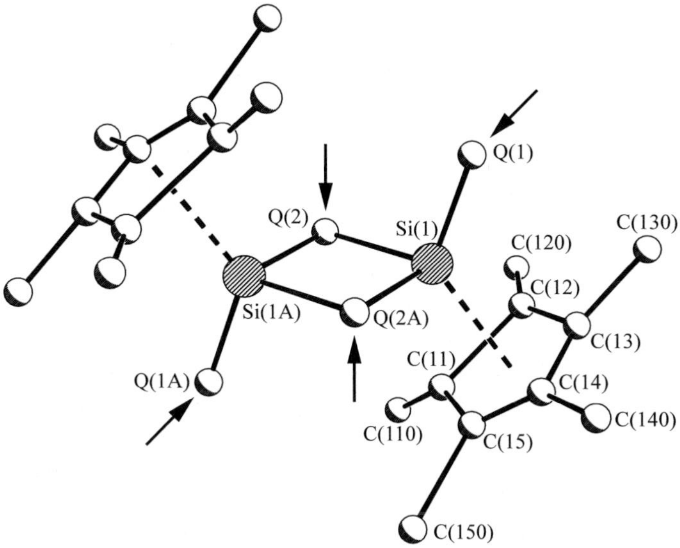

Fig. 4.6 Full molecule (asymmetric unit plus symmetry equivalents) of the model in the file si-01.res with the two highest residual electron density maxima and their symmetry equivalents. It is clear that Q(1) and Q(2) (see arrows) are indeed meaningful, even though they each represent only about two electrons in the current model.

will discover that even though the highest one is not very large, the two first peaks in the list are significantly above the others:

```
Q1    1   0.2617  0.1339 -0.1177  11.00000  0.05   1.99
Q2    1   0.0451 -0.1536 -0.0046  11.00000  0.05   1.65
Q3    1   0.6113  0.0080  0.3973  11.00000  0.05   0.47
Q4    1   0.1254  0.1026 -0.2352  11.00000  0.05   0.47
Q5    1   0.8217  0.7712  0.2816  11.00000  0.05   0.47
Q6    1   0.8779  0.6366  0.3773  11.00000  0.05   0.45
(...)
```

Figure 4.6 shows the full molecule of the current model with Q(1), Q(2) and their symmetry equivalents. From the position it is obvious that these two residual electron density maxima are indeed meaningful. However, which atom type could it be with only about two electrons?

A possible answer could lie within the model we already have: we know for sure that the ten carbon atoms are carbon atoms because of their geometry. There is no other ligand that looks like a Cp*, and yet they do not look right: the displacement parameters are dramatically inflated and their geometry is distorted. A possible reason for this could be partial occupancy, which is not realistic here. Another reason could be incorrect scaling. What if the entire scaling was wrong and the overall scale factor (*osf* or first free variable) overrated? This would result in a reduction of overall electron density and, in turn, in enlarged U_{eq} values for all atoms. The average U_{eq} value for the carbon atoms is 0.26, whereas one would expect about 0.05. What if

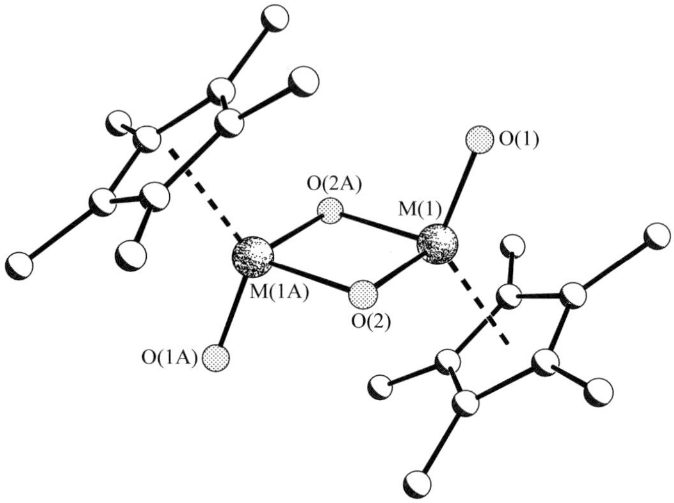

Fig. 4.7 Crystal structure of an unknown heavy metal compound (from a later stage of the refinement). In the file s-02.ins the metal is chosen to be Yb; the atoms labeled as oxygen could also be fluorine.

everything was off by roughly factor five? Multiplying all electron density by about five would lead to much more realistic ellipsoids for the carbon atoms and would make the two much too weak residual electron density peaks be between 8 and 10 electrons. That brings them in the range of oxygen or fluorine, two elements that form four-membered rings with metals and are also found as terminal ligands in the way shown in Figures 4.6 and 4.7. If this is true, however, it also means that the metal in our structure is about five times heavier than silicon[11]; somewhere between Yb and W perhaps. As a first attempt, we change the metal atom to Yb (do not forget the `SFAC` list) and assign oxygen to Q(1) and Q(2). This has been done in the file si-02.ins.

The first thing we notice after 20 refinement cycles is that the $R1$ value is much lower now—a good sign. The U_{eq} values for most of the atoms are reasonable and we have four residual electron density maxima that are significantly higher than the others. Two carbon atoms, however, C(12) and C(13), still have very large U_{eq} values. When you look at the structure with a graphical interface, it becomes visible that some of the carbon atoms have gone astray and two, C(12) and C(13), the two with the unreasonably high U_{eq} values, have totally wandered off. Q(1) and Q(4) have taken the place of carbon atoms, while Q(2) and Q(3) are found very close to the metal atom. It is a good idea to rename most of the carbon atoms (to retain a proper naming scheme), delete C(12) and C(13) and include Q(1) and Q(4) into the model as the new C(12) and C(13). Q(2), Q(3), and the lower residual electron density peaks can be ignored for the moment. Figure 4.8 shows the location of the atoms in si-02.res; all changes described have been done in si-03.ins.

[11] This is a big relief as Si is not exactly famous for carrying Cp* ligands.

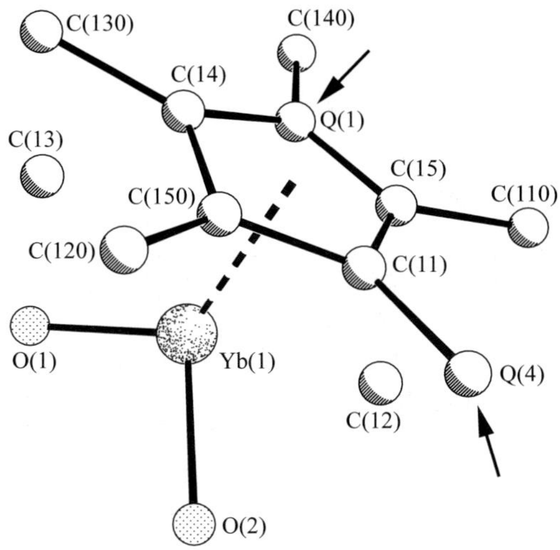

Fig. 4.8 Model as in the file si-02.res (asymmetric unit only). The carbon atoms C(12) and C(13) have wondered off, some other carbon atoms have swapped places and the residual electron density maxima Q(1) and Q(4) (see arrows) have taken the place of two carbon atoms. The overall geometry of the Cp* ligand is much better now than in sin-01.res (Figure 4.6).

After 10 cycles of refinement with SHELXL, the model has again improved: the *R* factors are lower than before and all U_{eq} values are within a reasonable range. On the other hand, the U_{eq} value for Yb(1) is still a little too low. This finding, together with the two very high residual density peaks (Q(1) and Q(2)) on either side of the metal, make it advisable to try other, even heavier metals. First, however, we should try to refine the model anisotropically. In the presence of much heavier atoms next to lighter ones, it is usually advisable to allow anisotropic refinement for the heavy atoms first. Therefore edit the file si-03.res by adding the line `ANIS $Yb` and save it as si-04.ins.

Not surprisingly the *R*-factors improve much. Now allow all atoms to be refined anisotropically by adding `ANIS` to the file si-04.res and renaming it si-05.ins. The results of the last round of refinement confront us with a new problem: one of the carbon atoms, C(130), has become ‘non-positive definite’ or NPD. This means that one or more elements of the anisotropic displacement parameter have become negative, which is physically meaningless. There are several possible reasons for an atom to become NPD, and one is incorrect scaling. A quick fix in the present situation is to constrain all anisotropic displacement parameters of the methyl groups to be identical. This can be achieved by adding the line

```
EADP C110 C120 C130 C140 C150
```

into the .ins file and save it as si-06.ins.[12]

[12] `EADP` stands for ‘equal anisotropic displacement parameters’; see also Section 2.5.6.

Now we can vary the atom type of the metal to see which element gives the best $R1$-value. This has been done for the elements Yb to Au, corresponding to the files si-06.ins to si-15.ins. The table below shows the $R1$-values as a function of the metal atom in the model.

Metal	Yb	Lu	Hf	Ta	W	Re	Os	Ir	Pt	Au
$R1$ [%]	4.63	4.61	4.60	4.60	4.60	4.60	4.60	4.61	4.63	4.64

This curve has a flat minimum around W, but this alone is by no means proof that the metal is indeed tungsten. We also have yet to establish the nature of the two atoms currently refined as oxygen, which could be nitrogen, oxygen or maybe even fluorine. Frequently, a search of the Cambridge Structure Database (CSD, Allen 2002) can shed some light on situations like this.[13] Searching for four membered M_2X_2 rings, where M is any metal between Yb and Au and X is N, O or F, resulted in 84 hits, 30 of which were for M = Re. The second largest group was for M = W with 12 hits. Other metals appear less likely. In all of the Re-structures, X was oxygen. Except for one case with a W–N double bond, X was O or F in the all of the tungsten structures. As the $M–X_{endo}$ distance is too long for a W–N double bond, the endocyclic non-metal atom is most likely oxygen or (for M = W) maybe fluorine. Taking into account that two of the Re-structures found in the CSD are very similar to our case (the monohydrate of $(ReO_2)_2Cp^*{}_2$ and one molecule with methyl groups instead of the Cp* rings)[14] and that there are no such tungsten containing molecules, we can assume with some probability that the central four-membered ring consists of rhenium and oxygen.

Judging from the $M–X_{exo}$ distance (1.71 Å), the terminal atom is most probably also oxygen. The final model, $(ReO_2)_2Cp^*{}_2$, as shown in Figure 4.9 (corresponding to si-17.res), is chemically reasonable, the U_{eq} values and figures of merit are sensible and the bond lengths are in agreement with comparable values found in the database. We do not, however, have any proof at all and the fact that the `EADP` constraint cannot be removed without making one of the carbon atoms become NPD is not encouraging. Other, for example spectroscopic methods could have helped, however the one crystal used for structure determination was the only sample of this molecule that had ever been found and the amounts were insufficient for further analyses.

Be the metal what it may, a question that remains is 'where did the crystal come from'? The flask was supposed to contain a Si–N compound, with no other atoms besides Si, N, C, and H. Nobody in the lab from which the sample originated uses Cp* ligands or any heavy metals, not even in the form of catalysts (it would not have been the first time that a catalyst or a degradation product of the catalyst had crystallized instead of the real product). The most plausible explanation is that the crystal was stuck to the spatula before it was ever inserted into the flask containing the Si–N compound. This would also explain why there was no second crystal to

[13] Even though the Cambridge Structure Database—distributed by the Cambridge Crystallographic Data Centre—is not free of charge, it is an invaluable tool for any crystallographic facility and no serious X-ray lab should be without it.

[14] The CSD codes for those two structures are GIPXAE and SUTHOE.

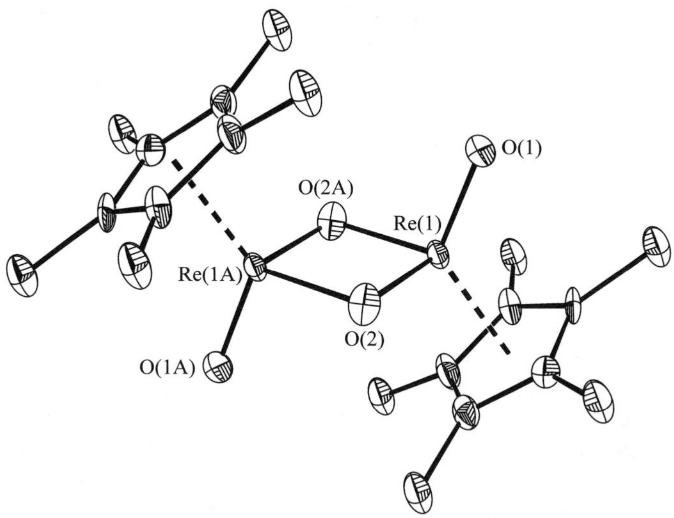

Fig. 4.9 Final anisotropic model of the unknown heavy metal complex (as in the file si-17.res). The metal has been refined as rhenium, the endocyclic and exocyclic atoms as oxygen. Hydrogen atoms have been omitted for clarity.

be found in the flask. Talking to people that had used the diffractometer in the days before this occurrence did not reveal the true nature or origin of the crystal, nor did extensive questioning of virtually everybody in the institute. Another mystery unsolved.

5

Disorder

A crystal is a potentially endless, three-dimensional, periodic discontinuum built up by atoms, ions or molecules. Because of the periodicity, every object is regularly repeated in three-dimensional space; that means every unit cell has exactly the same orientation with all molecules in the same conformation as in the cells to its left, right, top, bottom, front, and back. However, an *ideal crystal* does not exist; in most *real crystals* there are several lattice defects and/or impurities. Frequently, parts of molecules (or in some extreme cases whole molecules) are found in more than one crystallographically independent orientation.[1] One can distinguish three cases:

(1) more than one molecule per asymmetric unit
(2) twinning
(3) disorder

In the case of disorder, the orientations of some atoms differ randomly in the different unit cells. Picture a thousand soldiers lined up neatly in a square, who are supposed to all look to the right, but some 20 percent of them misunderstood the order and turn their heads to the left. This arrangement would be much like a two-dimensional crystal with an 80:20 disorder.

The structure determined from the diffraction pattern is the *spatial average* over the whole crystal. In by far most cases, disorder only affects small parts of molecules like organic side chains or $SiMe_3$-groups, or the heads of the soldiers from the above example. Another typical case is the *tert*-butyl group, which is normally almost free to rotate. Disorders of free solvent molecules located in holes in the crystal lattice are also very common. In principle, the presence of disorder is in contradiction to the definition of the crystalline state given above. Yet normally the order predominates, especially when only two different conformations are present in the crystal. Therefore, the conditions for X-ray diffraction are fulfilled, and the diffraction pattern looks unsuspicious. Normally, the solution and initial refinement of a partially disordered structure are not problematic. However, the ellipsoids derived from the anisotropic displacement parameters (ADPs) may be of pathological shape because the program tries to describe two or more atom sites with only one ellipsoid (see Figure 5.1), and the presence of relatively high residual electron density peaks or holes close to the disordered atoms is not unusual.

It has been shown that some disorders vanish below certain temperatures. This proves that disorder is not necessarily static: in such cases the observed disorder

[1] Owing to space group symmetry, the molecules forming a crystal always possess more than one orientation (except for the space group $P1$). This, of course, is not a disorder.

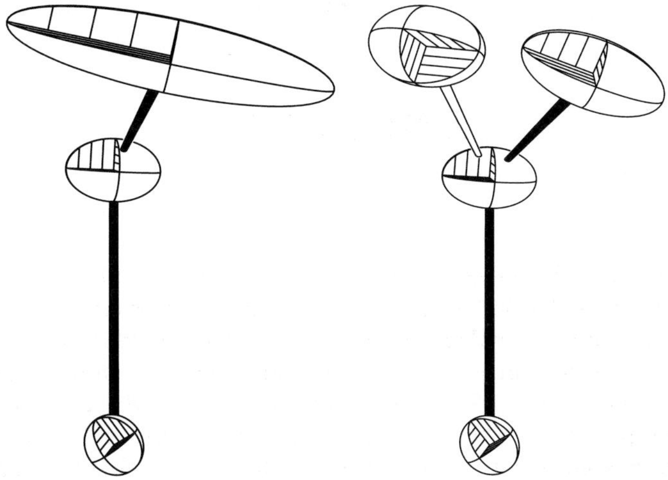

Fig. 5.1 Anisotropic displacement parameters of a disordered ethyl group on the left without and on the right with modeling of the disorder (empty lines for the minor component). If the disorder is ignored, the refinement program tries to describe both atom positions with one ellipsoid, giving rise to a cigar-shaped probability ellipsoid.

can be understood as real movement in the crystal, a movement that can be stopped by freezing the crystal. Therefore, besides other advantages of cryocrystallography, collecting data at low temperature can help avoid or reduce disorder. Other disorders show no temperature dependence and one can postulate the existence of two (or occasionally more) different types of unit cells; the disorder arose during crystal growth. In such a case it might help to grow the crystals at a lower temperature to reduce the likelihood of disorder. At lower temperatures the relative differences between two energetically similar orientations would become more distinct, which could lead to a stronger preference for one of them. Crystals also frequently grow more slowly and nicely at lower temperatures (provided one compensates for the temperature dependence of the solubility). In some cases, freezing of the crystal can convert dynamic to static disorder.

5.1 Types of disorder

In principle one can distinguish two types of disorder:

(1) Substitutional disorder
(2) Discrete (static) or continuous (dynamic) positional disorder

5.1.1 Substitutional disorder

Substitutional disorder describes a situation in which the same site in two unit cells is occupied by different types of atoms. This type of disorder is especially known from minerals and salt-like crystals: in some zeolites the Al- and Si-atoms share the same sites. In biological structures sometimes water molecules share a site with sodium, chloride or other ions. The refinement of substitutional disorder is relatively easy. Nevertheless one should know about it and should be able to recognize substitutional disorder. The clearest warning sign (in most cases the only one) are too small or too large anisotropic (or isotropic) displacement parameters.

Partial occupancy of atom sites is a relatively common special case of substitutional disorder, and non-coordinating solvent molecules are frequently found to occupy only about half of the voids in the crystal lattice. The presence of 'half waters' in protein structures is a typical example. Unusually high displacement parameters are a sign for partially occupied solvent molecules; however, one should take into account that, due to their mobility, even fully occupied non-coordinating solvent molecules tend to show relatively high displacement parameters. Therefore, the ADPs should be drastically larger to justify a reduction of the occupancy factors. The residual electron density map, which shows negative electron density at or around the nuclear positions if the true occupancy is lower than one, is a better criterion.

In some cases, mixed crystals can be treated as positional disorders if two similar molecules crystallize together at a single site in the same unit cell (see Example 5.3.3).

5.1.2 Positional disorder

Positional disorder is the 'normal case' of disorder: one atom occupies more than a single site. This can happen in a single unit cell (dynamic disorder, a real motion in the solid state) or distributed among different unit cells (static disorder, a look-alike motion). Both dynamic and static disorders are treated in the same way during refinement.

In a case of discrete disorder, the molecule can possess two (seldom more) well-defined energetically similar conformations. The example with the soldiers, some looking to the left and the others to the right, is such a case. In the spatial average—that is in the structure to be refined—one sees a superposition of both 'conformations'. The two positions appear as split atomic sites. Once recognized, such a disorder is refined relatively easily, as we will see below (Examples 5.3.1 and 5.3.2).

Continuous disorder is much more annoying (the soldiers from the example would all be shaking their heads, unable to decide whether to look to the left or to the right). If every rotational angle of for example a *tert*-Butyl group is energetically similar and there are no steric hindrances, this group of atoms might rotate virtually freely in the crystal (at least at room temperature), and in the spatial average one sees this group as a rotational toroid. It is hard to describe this situation to the refinement program. Normally, one reduces the problem to a refinement of only two or three sites per atom and accepts elongated ADPs (see Example 5.3.5). Fortunately, in many cases continuous disorder can be avoided or at least reduced by collecting low temperature data.

5.1.3 Mess—a special case of disorder

Especially in protein crystals, but also in other structures that show relatively large voids or cavities, one can find solvent molecules that do not show any order at all. Such solvent regions can be interpreted as a liquid that is amorphously frozen during data collection (assuming that you collect the data at low temperature as you always should). Following the Babinet Principle, this extreme case of disorder is described as *bulk solvent* and can be refined using a two-parameter approximation (Moews and Kretsinger, 1975). See Example 5.3.5.

5.2 Refinement of disorder

In most real-life cases it is sufficient to describe a disorder by formulating two different positions per disordered atom, and most refinement programs will not allow more than that anyway. The principle of disorder refinement is simple. The program needs to know the two sets of coordinates (that is positions) for each atom together with the relative occupancies (that is the ratio). For the case of our soldiers, by now already a little strained, it would suffice to say that 80% are looking to the right, while all the others have turned their heads to the left to describe the situation. The relative occupancies can either be given or refined; the latter is not possible with all refinement programs.

Finding the coordinates for both components is frequently much more complicated than formulating the disorder. In this context it is always a good idea to refine disorders at first *isotropically*, as anisotropic displacement parameters tend to compensate for the disorder, which makes it difficult to find additional positions (see Figure 5.1).

5.2.1 Refinement of disorder with SHELXL

SHELXL refines disorder by dividing the disordered atoms into groups, the components of the disorder. The occupancies of disordered groups are allowed to be refined freely. Together with the `PART` instruction, the introduction of so-called *free variables*[2] makes the refinement of disorders both easy and universal. For a better understanding, the refinement of positional disorder with only two components will be described. The refinement of disorders over several positions is done similarly.

The `PART` instruction

To begin with, the `PART` instruction in the .ins file divides the disordered atoms into two (or more) groups. Thus, each group represents one component of the disorder, and both groups contain the same atoms but on different sites. Practically, in front of the first disordered atom one writes `PART 1`, directly followed by all atoms of the first component. Directly before the atoms of the second component one writes

[2] Free variables are used in SHELXL for many other purposes in addition to describing disorders. For a detailed description of the general use of free variables, see Chapter 2.

`PART 2`. After all disordered atoms one writes `PART 0` to end the area of split sites. It is always a good idea to make sure that in both parts the atoms are in the same order. This enhances clarity and allows the use of `SAME` (see below).

The second free variable

In the next step, the ratio of the two components has to be taken into account. If the disorder does not involve any special positions, the occupancies of both components are allowed to possess any ratio. It is important, however, that the site occupancy factors (*sof*) sum up to exactly one.

The occupancy is refined with the help of a free variable, given in the .ins file. The line which directly precedes the first atom starts with `FVAR` and contains the overall scale factor (*osf*), also known as first free variable. For the refinement of a disorder, the *osf* should be followed by a second free variable whose value is between 0 and 1, describing the fraction of unit cells in the crystal showing the conformation described under `PART 1`. This means the second free variable is equivalent to the occupancy of the atoms in component one. For example a value of 0.6 for the second free variable corresponds to a ratio of 0.6:0.4, describing a 60–40% disorder. The values of the free variables are refined, but one must guess the initial value or estimate it from the peak height in the difference Fourier map. When in doubt, 0.6 is almost always a reasonable starting value.

Note that the refined value of any free variable has a calculated standard uncertainty, which is listed in the .lst file. The value of this standard uncertainty is supposed to be much smaller than the value for the free variable, or the disorder represented by the free variable would not be very meaningful.[3]

The site occupancy factor (`sof`)

Finally, the site occupancy factors of the disordered atoms must be manipulated to refer to the second free variable. This is done by changing the value of the *sof* instruction from `11.0000` to `21.0000` for the atoms in `PART 1` and to `−21.0000` for the `PART 2` atoms. The *sof* is given for each atom in the sixth column of the .ins file. `21.0000` means that the *sof* is set to '*1.0000 times the value of the second free variable*', while `−21.0000` sets the *sof* to '*one minus the value of the second free fariable*', completing the disorder.[4] Thus, the *sof* of both components add up to exactly 1.0000, while the ratio can be refined freely. The following example shows excerpts from an .ins file, describing the disorder of two carbon atoms. It was assumed that the component represented by `PART 1` would possess an occupancy of about 60%. The use of the `PART` instruction, the second free variable and the change of the *sof* instruction are highlighted in boldface font.

[3] If the free variable coupled to a disorder should refine to say 0.95 ± 0.1—this corresponds to an occupancy of the minor compound of $5 \pm 10\%$—we can assume that there is no disorder represented by the coordinates coupled to the free variable in question. In such a case, the atoms from the second component should be deleted, the *sof* instruction of the atoms of the first component should be changed back to `11.0000` and the `PART` instructions should be removed.

[4] This is one example for the description of a parameter (here the second free variable) as $10 \cdot m + p$. For a general description of this concept see Section 2.7.

```
FVAR     0.11272  0.6
(...)
PART 1
C1A   1  0.255905  0.173582  -0.001344   21.00000   0.05
C2A   1  0.125329  0.174477   0.044941   21.00000   0.05
PART 2
C1B   1  0.299373  0.178166  -0.015708  -21.00000   0.05
C2B   1  0.429867  0.176177  -0.062050  -21.00000   0.05
PART 0
```

Sometimes, an atom lies on a special position, and therefore its occupancy for the ordered case has to be reduced to 0.5 (e.g. an atom on a twofold axis or on an inversion centre) or 0.25 (atom on a fourfold axis), which corresponds to a *sof* instruction of `10.5000` or `10.2500`, respectively.[5] If such an atom is involved in a disorder, the *sof* instruction has to be changed to say `20.5000` or `20.2500` instead of `21.0000`. Later in this chapter the problem of disorders involving special positions will be discussed in detail.

How to find the second site

Disorder may be obvious if a second set of peaks appears in the difference Fourier map, or subtle if the ellipsoids stretch. If the ADPs of an atom behave strongly anisotropically, SHELXL writes a suggestion for the two possible sites of this atom into the .lst file. This message is to be found in the list of 'Principal mean square atomic displacements U' (located in the .lst file after the *R* value calculation following the last least squares cycle and right before the *K*-factor analysis and the list of most disagreeable reflections). However, not all 'may be split' atoms should be split; sometimes the anisotropic motion of an atom on a single position is a better description. When the positions of the disordered atoms are too far from each other to allow one ellipsoid to cover both sites, the anisotropy of the ADPs may not be strong enough for SHELXL to generate the message. In such cases, one can frequently use the coordinates of residual electron density peaks for the second site, or sometimes for both sites. These peaks are named Q by SHELXL and can be found at the very bottom of the .res file.

Sometimes, one does not see residual electron density near an atom and the ADPs are suspiciously elongated but not anisotropic enough for SHELXL to generate the 'may be split' message. In such a case one can use the same initial coordinates for both sites of a split atom; SHELXL separates them during the refinement. However, it is helpful to slightly 'move' one of the two sites by hand to circumvent a mathematical singularity.

[5] Other special positions or the combination of several special positions can lead to even lower occupancies. SHELXL recognizes atoms on or very close to special positions and automatically generates the constraints for all special positions in all space groups, which includes the reduction of the *sof*.

Disorder about special positions

If a molecule lies on a special position of higher symmetry than the molecule can possess, there are only two possibilities to eliminate this geometrical problem: either one changes the space group to one of lower symmetry without this particular special position, or—in most cases far better—one assumes a disorder of the molecule about this particular special position. A typical example is a toluene molecule on an inversion centre: none of the atoms lie on the special position, nevertheless in the *spatial average* the whole molecule is disordered in a ratio of 0.5 to 0.5 about the inversion centre.

The refinement of such disorders is relatively easy: the second site of each atom can be calculated directly from the positions of the atoms of the first component by means of the symmetry operator of the special position. Therefore, it is not necessary to have two parts in the .ins file. Instead of `PART 1`, `PART 2`, and `PART 0`, the disordered atoms are flanked with `PART -1` and `PART 0`. The negative part number suppresses the generation of special position constraints, and bonds to symmetry-related atoms are excluded from the connectivity table. Moreover, the use of the second free variable is not indicated in such a case, as the ratio between the components is determined by the multiplicity of the special position.

The site occupancy factors must take into account the multiplicity of the special position. For example, in the case of a mirror plane, a twofold axis and an inversion centre, the *sof* instruction has to possess the value `10.5000`. A threefold axis causes a *sof* instruction of `10.3333` and a fourfold axis one of `10.2500`, and so forth. SHELXL generates these site occupancy factors automatically only for atoms on or very close to special positions, but not necessarily for all atoms involved in a disorder about a special position.

Molecules that are located very close to special positions, so that the symmetry would lead to chemically unreasonable arrangements, are treated the same way. In such a case the `SPEC` instruction, which generates all appropriate special position constraints for the specified atoms, may be helpful too.

Disorders with more than two components

In some rare cases it can be appropriate to refine three components of a disorder. The atoms of the three components are grouped in `PART 1`, `PART 2`, and `PART 3`, and each component is associated with its own free variable, for example the free variables number 2, 3, and 4 (the first free variable is always the overall scale factor). Accordingly, the *sof* instructions need to be changed to `21.0000`, `31.0000`, and `41.0000`, and the sum of the three free variables must be one. With the help of the `SUMP` instruction, SHELXL combines free variables in the following way: the weighted sum of the specified free variables is restrained to possess a certain target value within a given standard deviation. Both the target value and the standard deviation can be chosen freely. In the case of a three component disorder associated to the free variables one, two, and three, the correct `SUMP` instruction is the following:

```
SUMP 1.0 0.001 1.0 2 1.0 3 1.0 4
```

Right after the `SUMP` command, the target value is given (1.0, as the three components must add up to precisely one), followed by the standard deviation (0.001). Thereafter, one finds pairs of weighting factors (here the weighting factors are all 1.0) with the numbers of the free variables (2, 3, and 4).

More than one disorder in a structure

If there are more than one independent disorders in a structure, one also has to use more then one additional free variable. Accordingly, the *sof* instructions are to be changed to `21.0000` and `-21.0000`, `31.0000` and `-31.0000`, `41.0000` and `-41.0000`, and so forth. For each disorder one uses `PART 1`, `PART 2`, and `PART 0`. Higher part numbers are only used to formulate disorders with more than one component. The format of the .ins file limits the number of free variables to 99, which should be enough to describe even very complicated structures.

Bulk solvent correction

SHELXL handles entirely disordered solvent with its bulk solvent correction. Using the algorithm of Moews and Kretsinger (1975), SHELXL refines two scaling parameters to describe bulk solvent: the first grows with the fraction of bulk solvent and usually possesses values of about 1. A high value of the second parameter means that only the very low resolution data is affected by the diffuse scattering of the bulk solvent (typical values are between 3 and 5). To apply the bulk solvent correction, one simply needs to add the command `SWAT` to the .ins file. SHELXL chooses starting values for the two parameters and refines them.

Disorder and restraints

Introduction of disorder into a model can increase the number of refined parameters quite considerably. In addition there frequently is a high correlation among parameters of disordered atoms (check the list of 'largest correlation matrix elements' in the .lst file). Thus, the refinement of disorders should always include restraints. Restraints are treated like experimental observations in the refinement (see Equation 2.6) and provide target values for particular parameters or link certain parameters, allowing the crystallographer to introduce chemical and physical information derived from sources other than the diffraction experiment into the refinement process. The following paragraphs give an overview of the restraints commonly used in connection with the refinement of disorder. A longer and more general description about restraints and how to use them can be found in Chapter 2. The following pages describe restraints only in their relevance for the refinement of disorders.

Similarity restraints

Equivalent bond lengths and angles in the two (or more) components of a disorder are assumed to be equal. If the atoms are in the same order in both components of a disorder, one may use the `SAME` instruction. The command `SAME`, followed by a list of atom names, must be located at the correct position within the .ins file. A `SAME` instruction makes the first atom in the list of atom names equivalent to the first atom immediately following the SAME command, the second atom equivalent to the second following, and so forth. 'Equivalent' means here that the 1,2- and 1,3-distances of corresponding atoms are restrained to be equal within certain standard deviations (default values are 0.02 for 1,2- and 0.04 for 1,3-distances). The program automatically sets up the $n \cdot (n-1)/2$ restraint equations that are required when n atoms should be equal. For a disordered tetrahydrofuran (*thf*) molecule (see Figure 5.2 for the atomic numbering scheme) the .ins file would look as follows (the letters A and B in the atom names refer to the components of the disorder: A for atoms in PART 1, B for atoms in PART 2).

```
FVAR  ....  0.6
(...)
PART 1
SAME O1B C1B C2B C3B C4B
SAME O1A C4A C3A C2A C1A
O1A   4   ....   ....   ....   21.000
C1A   1   ....   ....   ....   21.000
C2A   1   ....   ....   ....   21.000
C3A   1   ....   ....   ....   21.000
C4A   1   ....   ....   ....   21.000
PART 2
O1B   4   ....   ....   ....  -21.000
C1B   1   ....   ....   ....  -21.000
C2B   1   ....   ....   ....  -21.000
C3B   1   ....   ....   ....  -21.000
C4B   1   ....   ....   ....  -21.000
PART 0
```

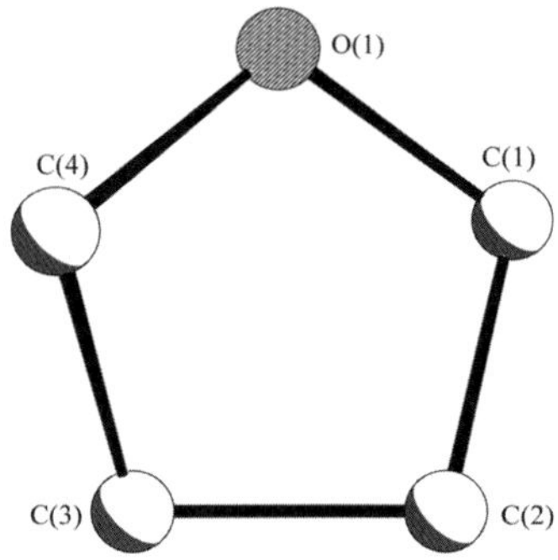

Fig. 5.2 Tetrahydrofurane molecule with atomic labeling scheme.

In this example the first SAME instruction precedes the atoms of the first component, listing the atoms of the second component. The oxygen atom O(1B), which is listed first after the word SAME, is made equivalent to O(1A), the first atom following the SAME command. Next, the carbon atom C(1B), the second atom listed in the SAME instruction is made equivalent to C(1A), the second atom following in the .ins file. Similarly, C(2B) is made equivalent to C(2A), C(3B) to C(3A), and C(4B) to C(4A), thus making the two components of the disorder equivalent. Thus all equivalent 1,2- and 1,3-distances between the two components are restrained to be the same. Thereby, in both components the atoms must be in the same order. The second SAME instruction also precedes the atoms of the first component, while the atoms that are listed with it are from the same (the first) component, but in backwards order. This assumes the oxygen atom O(1A) to be equivalent with itself, the carbon atom C(4A) equivalent to C(1A) of the same component, C(3A) equivalent with C(2A) and so forth, thus reflecting the symmetry within the *thf* molecule (see Figure 5.2). The combination of the two SAME instructions restrains equivalent 1,2- and 1,3-distances within each of the components *and* between the components to be the same. The second SAME instruction is not disorder specific but can also be used for *thf* molecules which are not disordered.

The list of atom names given in the SAME instruction may also contain the '<' or '>' symbols, meaning all intervening non-hydrogen atoms in a forward or backward direction, respectively. Thus, the two SAME commands in the example above could also have been formulated as follows:

```
SAME O1B > C4B
SAME O1A C4A < C1A
```

The SAME command is very powerful and by no means limited to the refinement of disorders. Whenever there is more than one molecule or group of atoms of the same kind in one structure (e.g. several *thf* molecules or $SiMe_3$ groups, or more than one molecule per asymmetric unit)—disordered or not—the SAME instruction efficiently restrains the bond lengths and angles to be similar. However, the SAME instruction is at the same time a sitting duck for mistakes. If the atoms in the two components (or independent molecules or groups of atoms) are not precisely in the same order, the restraints generated by the SAME command may do more harm than good. Typing errors in the list of atom names that follow the SAME command are also often fatal.

Alternatively to the SAME command, the distances between arbitrary atom pairs can be restrained to possess the same value using the SADI instruction. SADI is given together with a list of atom pairs. The distances between all pairs mentioned in a single SADI instruction are restrained to be equal within a certain standard deviation (default value is 0.02 Å). Restraining distances to a certain target value can be done using DFIX or DANG. Formulating exactly the same restraints for all equivalent distances with SADI as generated by the two SAME commands in the *thf* example above would require the following six lines:

```
SADI O1A C1A O1A C4A O1B C1B O1B C4B
SADI C1A C2A C3A C4A C1B C2B C3B C4B
SADI C2A C3A C2B C3B
SADI 0.04 O1A C2A O1A C3A O1B C1B O1B C3B
SADI 0.04 C1A C3A C2A C4A C1B C3B C2B C4B
SADI 0.04 C1A C4A C1B C4B
```

The 0.04 for the last three commands changes the standard uncertainty from the default value 0.02, which is suitable for 1,2-distances, to 0.04, a value more reasonable for 1,3-distances.

SIMU/DELU

Disordered atoms tend to show problems when the first attempts are made to refine them anisotropically. Figure 5.3 shows ellipsoids representing anisotropic displacement parameters and the effect of applying restraints to the ADPs. The similar-ADP restraint SIMU and the rigid-bond restraint DELU should be used in disorders to make the ADP values of the disordered atoms more reasonable. SIMU restrains the anisotropic displacement parameters of adjacent atoms to be similar, and DELU enforces that the main directions of movement of covalently bonded atoms are the same. The default values for the standard deviations are 0.04 for SIMU (0.08 for terminal atoms, which tend to move more strongly) and 0.01 for DELU. Note that SIMU is

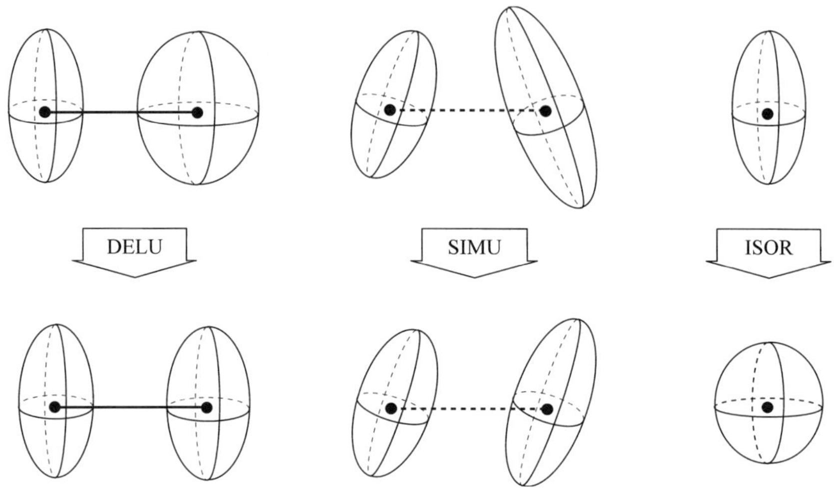

Fig. 5.3 Effect of the restraints DELU, SIMU, and ISOR. Illustration taken from Schneider (1996b). This is exactly the same as Figure 2.2.

only meaningful for anisotropically refined atoms and is ignored by SHELXL if the specified atoms are still isotropic.[6] `DELU`, in contrast, can be applied to isotropically refined atoms as well.

`ISOR`

In particularly tough cases of disorders, especially for disordered solvent molecules, it can be useful to restrain the anisotropic U^{ij}-values of the atoms to behave more isotropically with `ISOR`. As with `DELU` SHELXL ignores `ISOR` commands for atoms that are refined isotropically. `ISOR` is helpful for certain special cases (e.g. a disordered atom close to a special position, or anisotropic refinement of a protein against 1.4 Å data) and should almost always be applied to the water molecules of a protein model, but is otherwise less appropriate than `SIMU` or `DELU`. Figure 5.3 illustrates the effect the restraints `DELU`, `SIMU`, and `ISOR` have on anisotropic displacement parameters.

`FLAT`

If four or more atoms are supposed to lie on a common plane (e.g. atoms of an aromatic system) one can use `FLAT` to restrain them to do so within a given standard deviation (default value 0.1 Å^3).

Disorder and constraints

In some cases, even *constraints* can be used to refine disorders. As restraints, constraints improve the data to parameter ratio, however not by contributing

[6] There is no harm in using `SIMU` on atoms that are still refined isotropically. `SIMU` will become effective as soon as the atoms corresponding to the restraint are refined anisotropically.

observations but by decreasing the number of parameters to be refined. Constraints are exact mathematical relationships relating certain parameters and have no standard uncertainty.

EXYZ (for Equal XYZ) followed by a list of atom names forces the named atoms to possess the same coordinates as the first atom of the list. This can be useful for some types of substitutional disorder, for example a phosphate and a sulfate ion sharing the same site in a structure.

EADP (for Equal ADP) followed by a list of atom names makes the anisotropic displacement parameters of all named atoms equal to those of the first atom in the list.

If one encounters 'geometrical problems', say a phenyl ring does not want to be hexagonal, the **AFIX** constraints can help: `AFIX 66` forces the six following non-hydrogen atoms to form a regular hexagon, while `AFIX 56` defines a regular pentagon. This should be done with care and preferably only in early stages of a disorder refinement. Whenever a similarly satisfying effect can be reached by the use of restraints, the restraints should be given the preference.

General remarks

To make sure that the two sites of an atom are clearly separated and not fitted by an ellipsoid, it is necessary to make all disordered atoms isotropic (if they are not already) prior to the formulation of the disorder. Once the two positions seem to be stable, one can proceed to refine them anisotropically, preferably with the help of restraints.

In any case, a disorder must be chemically reasonable. Not every significant residual electron density peak is caused by disorder. High residual electron density can also be caused by inadequately corrected absorption, Fourier series truncation errors (for example when strong reflections are missing) or radiation damage. Such artefacts often lead to the accumulation of spurious electron density at special positions.

5.3 Examples

In the following sections I present examples of how to parameterize disorder for refinement with SHELXL. All files you may need in order to perform the refinements yourself are given on the CD-ROM that accompanies this book. The first example is an easy and straightforward case of static positional disorder that should acquaint you with the `PART` command, the free variables and overall scale factors, as well as the use of restraints and the addition of hydrogen atoms to disordered molecules. The second case is a very difficult static positional disorder that affects most of the molecules and involves a special position. You will learn from this example the use of restraints involving symmetry equivalents of atoms (using the `EQIV` command) and you will develop a thorough understanding of the phenomenon of disorder by refining this structure yourself step by step. The third example describes a mixed crystal, treated as substitutional disorder. The next four examples deal with disordered solvent molecules and should give you an idea of what these molecules can do in a structure. You will learn the use of the `PART -1` instruction for two molecules on twofold axes not fulfilling the symmetry of the special position. The last example in this chapter shows three different kinds of disorder appearing together in the same structure: dynamic positional disorder, solvent disorder and bulk solvent. This case introduces the `SWAT` command and demonstrates that it can be tedious to refine continuous disorder.

5.3.1 *Gallium-iminosilicate—Disorder of two ethyl groups*

The gallium-iminosilicate $[RSi(NH)_3GaEtGaEt_2]_2$ where R is 2,5-tPr$_2$C$_6$H$_3$NSi Me$_2$$i$Pr (see Figure 5.4), crystallizes in the monoclinic space group *C2/c* with half a molecule per asymmetric unit. The other half of the molecule is generated from the first one by the crystallographic inversion centre. The core of the molecule can be described as a bi-truncated square bi-pyramid. It is formed by four six-membered $SiGa_2N_3$-rings in the boat conformation, sharing the three stern-atoms and the three prow-atoms with each neighbour ring. Thus, two planar four-membered $SiGaN_2$-rings are formed. In all rings, metal and nitrogen atoms occupy alternating sites. This kind of cage is also known from tetraasteran ($C_{12}H_{16}$, see Hutmacher *et al.*, 1975). A more detailed description of the molecule and the chemistry behind it can be found in Rennekamp *et al.*, (2000).

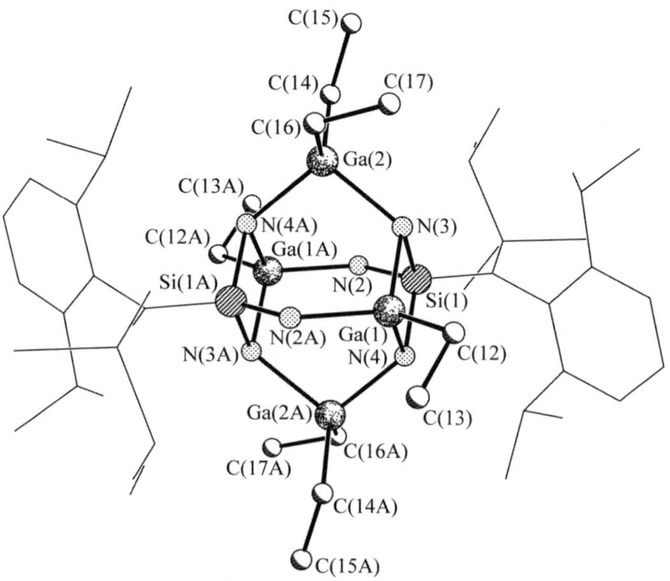

Fig. 5.4 Ball and stick representation of $[RSi(NH)_3GaEtGaEt_2]_2$. The 2,5-tPr$_2$C$_6$H$_3$NSiMe$_2$$i$Pr parts have been drawn in thin lines and all hydrogen atoms bound to carbon atoms have been omitted for clarity.

Ethyl groups frequently tend to be disordered. The above described gallium cage is a good example, as two of the three crystallographically independent ethyl groups—one per Ga-atom—show rotational disorder about the Ga-C-axis. The solution and first refinement steps of this structure are very straightforward, and we join the refinement at a point where the first signs of disorder appear. This is the file ga-01.res on the accompanying CD-ROM, which contains the complete isotropic model without hydrogen atoms. When you look at the atoms and difference density peaks in this file with a graphical interface such as XP or Ortep the following becomes visible: the highest residual electron density peaks are near the two Ga atoms: Q(1) (2.28 electrons per Å^3) and Q(2) (2.23 electrons) near Ga(1) and Q(7) (1.16 electrons) and Q(8) (1.10 electrons) close to Ga(2). This is a normal effect for isotropically refined heavy metals. In addition, relatively high residual density can be found near two of the three independent ethyl groups: Q(3), Q(4), and Q(5) with 2.20, 1.79, and 1.31 electrons, respectively. The latter three residual electron density maxima could indicate disorder of the two ethyl groups. However the high residual density close to the metal atoms reduces the significance of the other maxima. It should be a good idea to first refine all metal atoms (i.e. gallium and silicon) in the structure anisotropically, and then examine the remaining residual electron density. For that purpose we add the following instruction directly above the first atom of the atom list.

```
ANIS $GA $SI
```

This has been done in the file ga-02.ins (the file ga-01.res was renamed to ga-02.ins, after the changes were done). After ten cycles of refinement with SHELXL, the results are in the file ga-02.res. As shown in Figure 5.5, the atomic displacement parameters of the atoms C(13) and C(15) are somewhat too large. In addition, next to these atoms appear the three highest residual electron density maxima: Q(1), Q(2), and Q(3), with 2.17, 1.71, and 1.33 electrons per Å^3. This is typical for a disorder. The logical interpretation is: Q(1) is the second site of C(15), and Q(2) and Q(3) are two new positions for C(13), *replacing* the current C(13). The next highest density maximum, Q(4) (1.23 electrons) lies on an aromatic bond and is not part of a disorder. Q(5) and the other maxima are too weak to be relevant at this state of the refinement. To formulate the disorder, use the `PART` instruction, introduce two new free variables, change the *sof* instructions and make the following changes:

C(15) → C(15A)
Q(1) → C(15B)
Q(2) → C(13A)
Q(3) → C(13B)
Delete the old C(13)

As there is no disorder refinement without restraints, you should use `SAME` (or the respective `SADI` instructions) to make the 1,2- and 1,3-distances equivalent. To catch all possible 1,3-distances, start two atoms earlier. That means the `SAME` instructions should not be given immediately before the disordered atoms

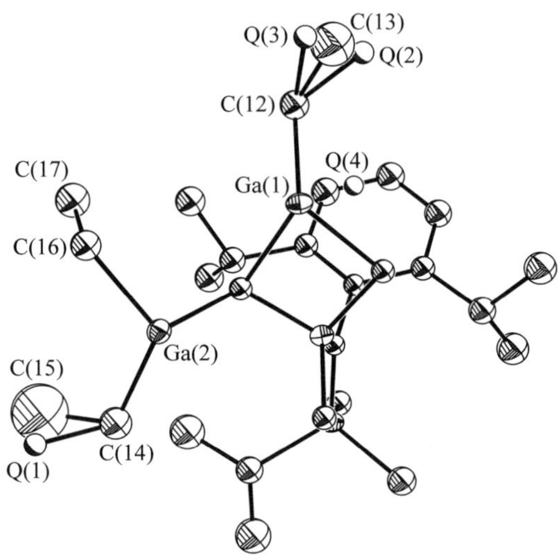

Fig. 5.5 Asymmetric unit for ga-02.res with the four highest residual electron density maxima. Only the metal atoms are refined anisotropically.

(C(13) and C(15)) but rather two atoms before them (right before Ga(1) and Ga(2)). Also make sure that the atoms are all in the right order! The similarity restraints `SIMU` and `DELU` (`DELU` is ignored by SHELXL if the named atoms are refined isotropically) make the atomic displacement parameters more reasonable. The critical portion of the new file, ga-03.ins, looks like this:

```
SIMU c12 c13a c13b c14 c15a c15b
DELU c12 c13a c13b c14 c15a c15b
WGHT      0.100000
FVAR      0.11272  0.6  0.6
same ga1 c12 c13b
GA1 5  0.447952  1.122706  0.039108  11.00000 0.01492 0.02158 =
       0.01663 -0.00262  0.00225  0.00363
C12 1  0.400303  1.237823  0.073859  11.00000 0.02906
PART 1
C13A  1  0.4379   1.3631    0.0949    21.00000 0.05
PART 2
C13B  1  0.4371   1.2955    0.1347   -21.00000 0.05
PART 0
same ga2 c14 c15b
GA2 5  0.445620  0.809823  0.031364  11.00000 0.01782 0.02043 =
       0.01730  0.00047  0.00362 -0.00096
C14 1  0.423155  0.663631 -0.020808  11.00000 0.03426
PART 1
C15A  1 0.375224 0.581908 -0.005250 31.00000 0.11712
PART 2
C15B  1 0.4151   0.5406    0.0044   -31.00000 0.05
PART 0
```

The next .lst file will contain all the information you need to check whether the restraints are used properly. If `MORE 3` is given in the .ins file the .lst file contains a list of all distances treated as equivalent by SHELXL. After ten cycles of SHELXL, the files ga-03.res and ga-03.lst contain the results of the disorder refinement (see Figure 5.6).

The highest residual electron density maximum (Q(1) with 1.21 electrons per Å^3) lies on an aromatic bond—where Q(4) was located before. Taking into account that the model is still mainly isotropic, the other electron density maxima are unsuspicious. In the next step we can refine all atoms anisotropically by including `ANIS` in the .ins file, directly before the first atom. To make sure that you can find all possible hydrogen sites in the difference density, give `PLAN 60` instead of `PLAN 20`. This makes SHELXL find 60 residual electron density peaks instead of 20. The file ga-04.ins contains all these changes.

Except for the H-atoms involved in the disorder, all hydrogen positions can be seen in the difference Fourier synthesis (see the Q-peaks in the file ga-04.res). To validate the restraints, examine the file ga-04.lst (especially the lines 137–175). The density peaks Q(12), Q(23), and Q(24) (0.65, 0.62, and 0.61 electrons) correspond to the hydrogen atoms bonded to N(2), N(3), and N(4), respectively. The following `HFIX` instructions cause SHELXL to geometrically calculate the hydrogen positions[7]:

`HFIX 43`	for all Ar-H
`HFIX 13`	for the CH-groups
`HFIX 23`	for the CH_2-groups (but **not** for C(12) and C(14); see below)
`HFIX 33`	for the disordered CH_3-groups
`HFIX 137`	for the other CH_3-groups

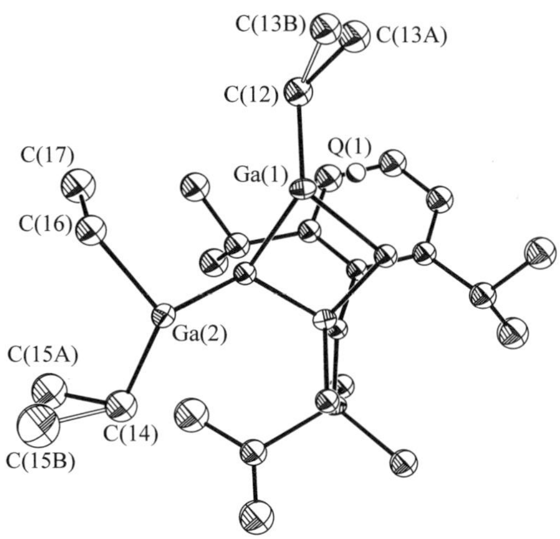

Fig. 5.6 Asymmetric unit for ga-03.res with refined disorder and highest residual electron density maximum. The structure is displayed in the same orientation as in Figure 5.5.

[7] A detailed introduction of the `HFIX` command is given in Section 3.3.

The H-atoms bonded to N(1), N(2), and N(3) are taken directly from the difference Fourier maxima Q(12), Q(23), and Q(24):

Q(12) ➔ H(2N)
Q(23) ➔ H(3N)
Q(24) ➔ H(4N)

The N–H distance is set to a value of 0.88 Å for N(2), which makes only two bonds to metal atoms, and to 0.91 Å for N(3) and N(4), which make three bonds to metal atoms, using the distance restraint DFIX:[8]

```
DFIX 0.88 N2 H2N
DFIX 0.91 N3 H3N N4 H4N
```

It is important to note that the positions of the hydrogen atoms bonded to C(12) and C(14) are disordered in the same way as the corresponding Me-groups, although C(12) and C(14) are not directly involved in the disorders themselves. To manage this problem, give after both C(12) and C(14) a `PART 1` and a `PART 2` instruction. In each part, write an `AFIX 23` and an `AFIX 0` instruction, flanking two H-atoms with the coordinates `0 0 0`. These coordinates are ignored by SHELXL, which calculates the correct positions following the geometry defined by the `AFIX` command.

```
C12   1 0.400238   1.237374   0.073651   11.00000 0.02185 0.03417 =
        0.03500  -0.01223  0.00818   0.00417
PART 1
AFIX 23
H12A 2 0   0   0    21.00   -1.200
H12B 2 0   0   0    21.00   -1.200
AFIX 0
PART 2
AFIX 23
H12C  2 0   0   0   -21.00   -1.200
H12D  2 0   0   0   -21.00   -1.200
AFIX 0
PART 1
C13A 1 0.440412   1.342682   0.107947   21.00000 0.04071 0.04595 =
        0.08781 -0.03921  0.00854  0.00640
PART 2
C13B 1 0.440475   1.309546   0.125832 -21.00000 0.03099 0.02988 =
        0.04538 -0.01776  0.02015  0.00523
PART 0
```

Finally, we can set PLAN back to 20. Everything described has been changed in the file ga-05.ins and after ten more cycles of refinement, we will have the files ga-05.res and ga-05.lst. The disordered CH_2 group at the C(12) site is shown in Figure 5.7.

[8] These distances are sensible at this temperature ($-140°C$). A list of X-H distances at the temperature defined by the TEMP instruction in the .ins file, can be found in the .lst file. See also Chapter 3.

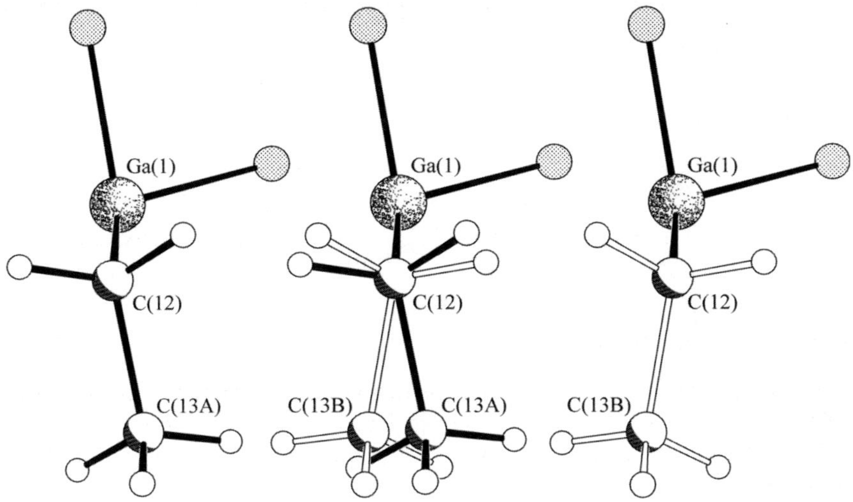

Fig. 5.7 Disordered CH_2 hydrogen atoms bonded to the non-disordered atom C(12). Left: major component (PART 1), middle: both components of the disorder, right: minor component (PART 2).

Look into the file ga-05.lst to verify that `HFIX 137` resulted in a defined torsion angle for all ethyl groups (the relevant part consists of lines 255–316). The file ga-05.res contains the final version of the refinement (see Figure 5.8).

Finally, the weighting scheme has to be refined to convergence. This has been done in the file ga-06.res, which represents the publishable final model.

5.3.2 Disorder of a Ti(III) compound

The Ti(III) complex $(\eta^5C_5Me_5)_2Ti_2(\mu\text{-}F)_8Al_4Me_8$ crystallizes in the monoclinic space group *C2/c* with half a molecule per asymmetric unit. The other half is generated from the first one by the crystallographic twofold axis through the atoms Al(1) and Al(3). The green crystals grow from toluene and are extremely sensitive to air: immediately after retaining them from the flask they start to decompose and lose the green colour. Only cooling the crystals under the microscope and low-temperature data collection made a structure determination possible. A more detailed description of the molecule and the chemistry behind it can be found in Yu *et al.*, (1999).

The problems with this structure start right at the beginning. The only information we have for model building is the following:

$Cp^*TiF_3 + 2\ AlMe_3$ ➔ green crystals[9]

[9] Cp* is pentamethylcyclopentadienyl

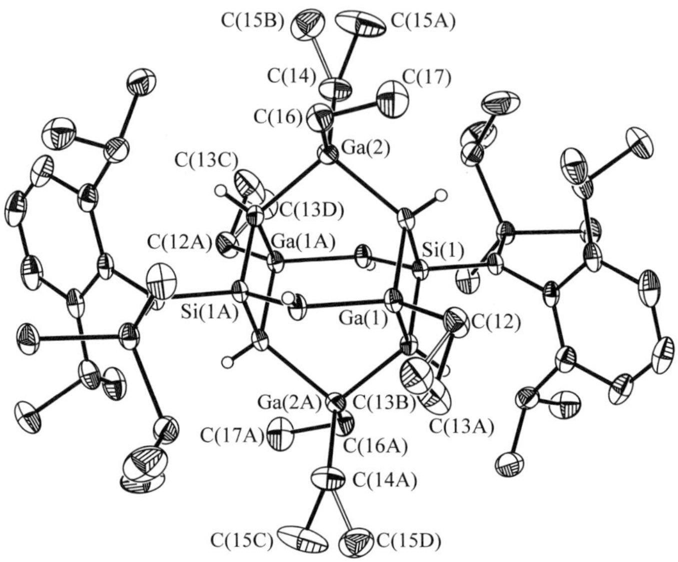

Fig. 5.8 Complete molecule, final model with the two disordered ethyl groups. Hydrogen atoms bonded to carbon atoms have been omitted for clarity; empty lines are used for bonds of the minor component.

The solution from SHELXS (ti-00.res) contains a titanium atom and a list of 39 unscaled electron density peaks. The peaks Q(24) to Q(39) are significantly weaker than Q(1) to Q(23) and can therefore be deleted. The remaining 24 atoms are arranged as shown in Figure 5.9. Clearly recognizable is the Cp* ring and the four F-atoms in form of the four highest residual maxima (Q(1), Q(2), Q(3), and Q(4)), as well as the titanium atom in the asymmetric unit (Figure 5.9, left). After generating the symmetry equivalents of the atoms in order to see the whole molecule, the rest of the Qs form a strange cage, which does not seem to make any chemical sense (Figure 5.9, right). These peaks are therefore deleted. The titanium and the fluorine atoms as well as the carbon atoms forming the Cp* ligand are retained in ti-01.ins (shown in Figure 5.10) and fed into SHELXL.

The five highest residual electron density maxima in the file ti-01.res (resulting from the first refinement job) are of about the same height (12.85 to 12.47 electrons per Å^3) and much higher then the other residual maxima (Q(6) has only 5.88 electrons). A peak height of about 13 electrons corresponds very well with aluminum, which is expected to bond to the fluorine atoms. However, the aluminum positions do not seem to be chemically reasonable (see Figure 5.11, left). Taking into account the symmetry equivalent atoms (labeled with an additional A after their original name), one sees the peaks Q(1), Q(2), Q(4), and Q(4A) form one component and the peaks Q(3), Q(3A), Q(5), and Q(5A) form the other one (see Figure 5.11, right). Q(1) and Q(2) both lie on the crystallographic twofold axis. To formulate the disorder, make the following changes: use the `PART` instruction, introduce a new free

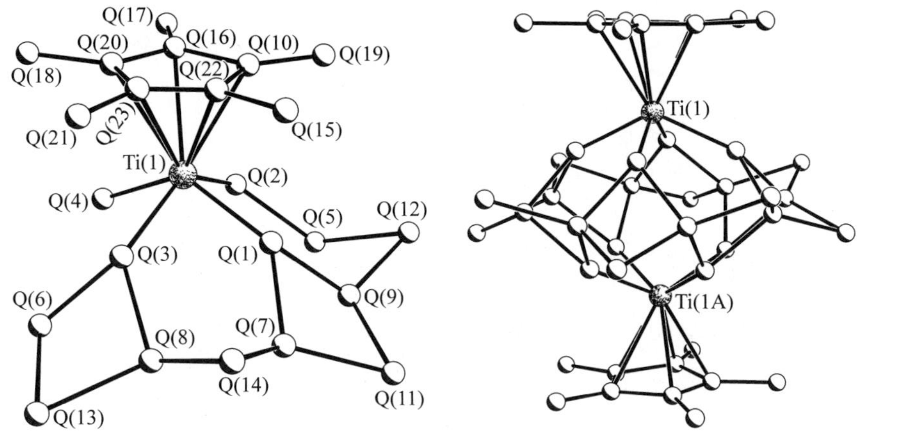

Fig. 5.9 Structure solution from SHELXS as in the file ti-00.res. Left: the asymmetric unit, right: the complete molecule.

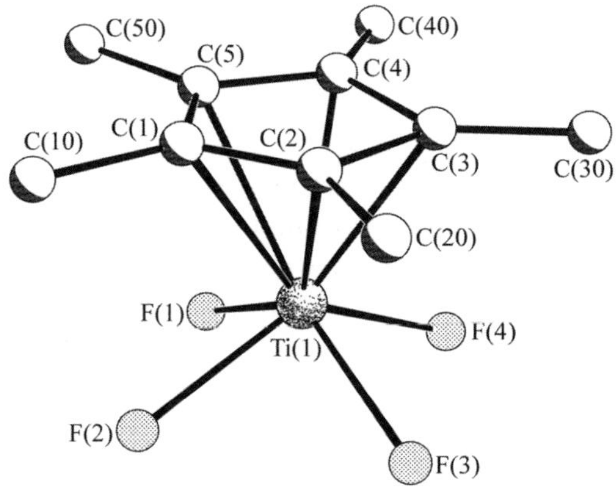

Fig. 5.10 Asymmetric unit as written to the file ti-01.ins.

variable, change the *sof* instructions and rename the residual electron density peaks Q(1) to Q(5) into aluminum atoms as follows:

Q(1) → Al(1) in `PART 1`
Q(2) → Al(2) in `PART 1`
Q(3) → Al(4) in `PART 2`
Q(4) → Al(3) in `PART 1`
Q(5) → Al(5) in `PART 2`

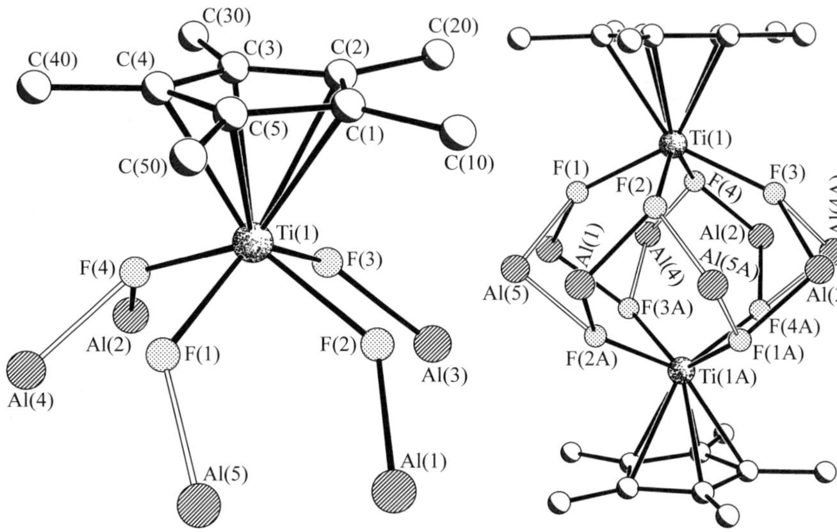

Fig. 5.11 New model as written to the file ti-02.ins. Left: asymmetric unit; right: complete molecule. Empty lines are used for bonds of atoms of the second component (Al(4) and Al(5) are in PART 2).

Note that the correct site occupancy factor instructions of Al(1) and Al(2) (both in `PART 1`) are `20.5000` and not `21.0000`, as they lie on the crystallographic twofold axis.

To restrain the U values of all Al atoms give `SIMU 0.04 0.08 2.5 $Al`; the $ sign means 'all'. The first two numbers after the `SIMU` command are the standard deviations for non-terminal and terminal atoms respectively (both the default values). The 2.5 is the radius of influence for the restraint. It is increased from its default value (1.7) as the Al-F distances could be slightly larger than 1.7 Å.

The refinement gives rise to the files ti-02.res and ti-02.lst and shows the following results: the second free variable refines to 0.51, which is a reasonable value. Q(1) to Q(4) (5.35 to 2.70 electrons) are significantly higher than the other residual electron density maxima, and seem to represent carbon atoms (see Figure 5.12). They bond to the disordered aluminum atoms, but are not disordered themselves (only their *connectivities* are disordered not their *sites*). Next step:

Q(1) → C(100)
Q(2) → C(200)
Q(3) → C(300)
Q(4) → C(400)

It is also a good idea to use some restraints on the Cp* ligand, which is assumed to posses D_{5h} symmetry. A single `SAME` command can restrain all 1,2- and 1,3-distances in the ligand to be identical (all together 25 restraints!).

```
same C2 > C5 C1 C20 > C50 C10
C1    1    0.193465    0.169566    0.113055    11.00000    0.02258
```

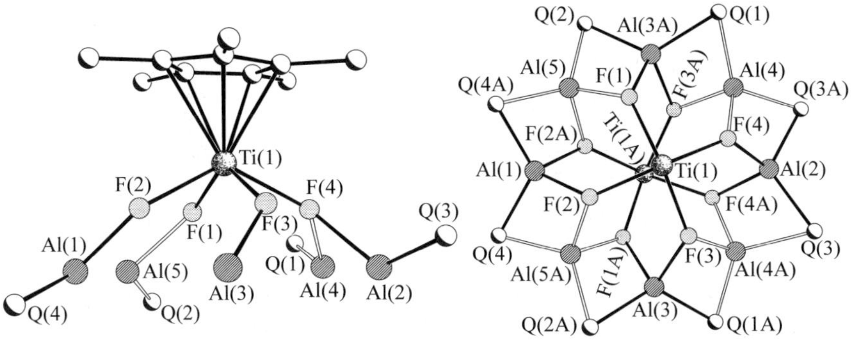

Fig. 5.12 The four highest residual electron density maxima from ti-02.res. Left: asymmetric unit; right: complete molecule. In the right picture the Cp* ligands have been omitted for clarity and the molecule is shown in top view.

```
C2    1    0.166825    0.249754    0.066287    11.00000    0.02231
C3    1    0.191162    0.324737    0.120094    11.00000    0.01964
C4    1    0.235954    0.291536    0.202630    11.00000    0.02627
C5    1    0.235499    0.197300    0.196759    11.00000    0.02431
C10   1    0.176273    0.077515    0.082855    11.00000    0.03727
C20   1    0.121213    0.255132   -0.024279    11.00000    0.03015
C30   1    0.181023    0.419057    0.098734    11.00000    0.03716
C40   1    0.276334    0.344412    0.275982    11.00000    0.04180
C50   1    0.269335    0.134130    0.264819    11.00000    0.04437
```

The first eight residual electron density maxima as found in ti-03.res are very close to the fluorine positions (see Figure 5.13). Together with the relatively high U values of the F-atoms, this result indicates that the fluorine atoms are also disordered. Therefore we delete all current F-atoms and replace them with the new sites taken from the Q-positions. To make sure that all new F-atoms belong to the right component, one should check the Al–F distances (or Al–Q distances, respectively), which are supposed to be about 1.7 Å. This is much easier after generating the symmetry equivalent atoms.

It becomes clear that Q(1) to Q(4) belong to `PART 1`, while Q(5) to Q(8) belong to the other component (do not forget to change the *sof* instructions to `21.0000` or `-21.0000` respectively, and the atom type number to 3).

Q(1) ➔ F(1A)
Q(2) ➔ F(2A)
Q(3) ➔ F(3A)
Q(4) ➔ F(4A)
Q(5) ➔ F(1B)
Q(6) ➔ F(2B)
Q(7) ➔ F(3B)
Q(8) ➔ F(4B)

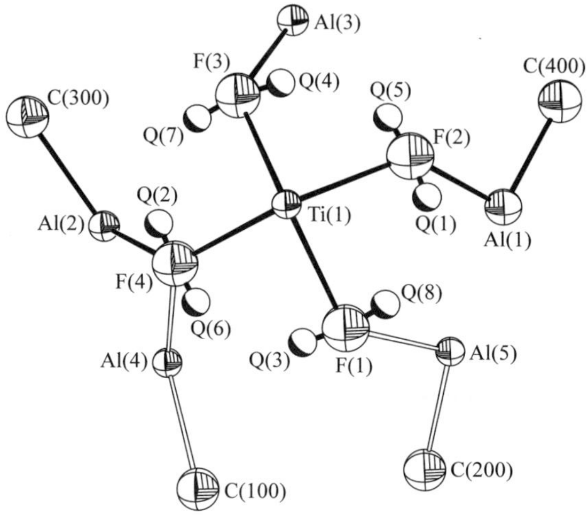

Fig. 5.13 Residual electron density maxima next to the fluorine positions indicating the F atoms to be part of the disorder.

Give also some similarity restraints for the disorder. `SIMU` and `DELU` should be used for the whole structure. In addition, equivalent 1,2- and 1,3 distances should be restrained to the same value using `SADI`. Be very careful when dealing with symmetry equivalents (use the `EQIV` instruction). All these changes have been made in the file ti-04.ins. The `EQIV/SADI` instructions look like this:

```
EQIV $1 -x, y, -z+1/2
SADI 0.04 Ti1 Al1 Ti1 Al2 Ti1 Al3 Ti1 Al4 Ti1 Al5
SADI F1A Al1 F2A Al2 F3A Al3_$1 F4A Al3 F1B Al5_$1 F2B Al4 F3B Al4_$1 F4B Al5
SADI C100 Al3_$1 C100 Al4 C200 Al5 C200 Al3_$1 C300 Al2 C300 Al4_$1 =
     C400 Al5_$1 C400 Al1
SADI F1A Ti1 F2A Ti1 F3A Ti1 F4A Ti1 F1A Ti1 F1B Ti1 F2B Ti1 F3B Ti1 F4B Ti1
```

After eight cycles of refinement, the *R*-values have already much improved and the highest residual electron density peak is at 0.99 electrons. Figure 5.14 shows the disorder in detail. In the next step, we can refine all atoms anisotropically by writing `ANIS` right before the first atom in ti-05.ins.

The hydrogen positions for the Cp* ligand can be found in the difference Fourier synthesis (see the Q-peaks in the file ti-05.res). Now we can add `HFIX 137` for the Cp* Me-groups. The hydrogen atoms of the Al–CH_3 groups are to be treated like the disordered CH_2-groups in the Ga-Iminosilicate (first example of this chapter, Figure 5.7). To avoid problems during the generation of the disordered hydrogen positions, which can occur in this rare case, it is useful to have all atoms of the model in the same asymmetric unit. Therefore, for the file ti-06.ins the coordinates of some atoms are transformed, and the distance restraints are changed accordingly. It is a little tedious to actually generate the symmetry equivalent atoms and change all the

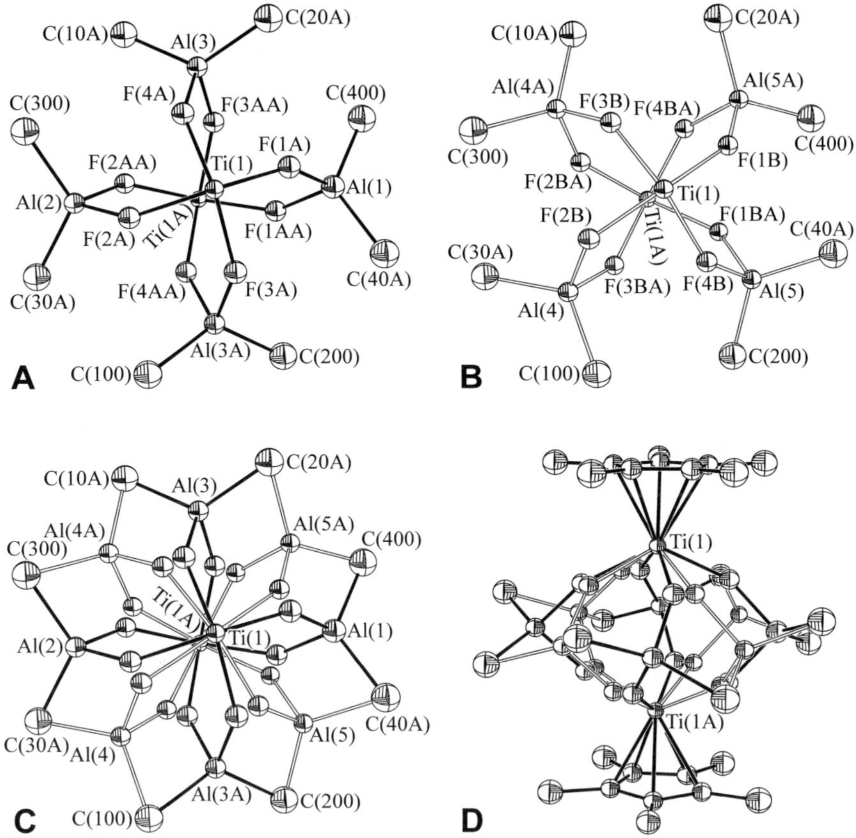

Fig. 5.14 Disorder as in ti-04.res. A: component 1, B: component 2, C: both components (in all three cases top view without the Cp* ligands), and D: side view to both components with the Cp* ligands.

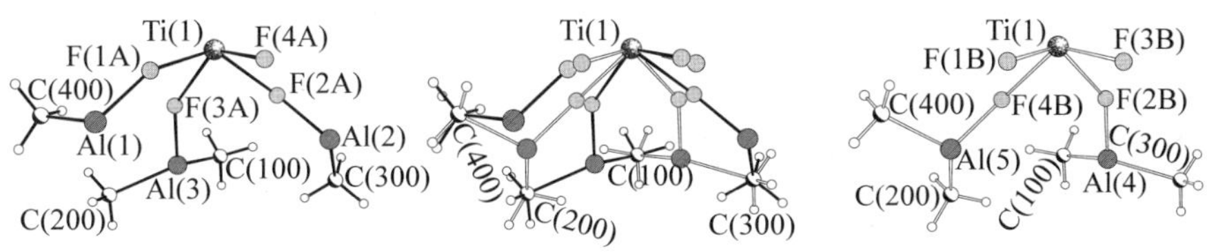

Fig. 5.15 Disorder of the hydrogen atoms. On the left side are the atoms of component one, on the right side the atoms of component two and in the middle both components.

restraints correctly. It is, however, very educational and explicitly recommended.[10] The disordered hydrogen positions are shown in Figure 5.15.

The file ti-06.res contains the complete anisotropic model with all hydrogen atoms. Finally, the weighting scheme has to be refined to convergence. This has been done in the file ti-07.res, and Figure 5.16 shows the final publishable model.

[10] If you are using XP, the commands ENVI and SGEN will prove very helpful.

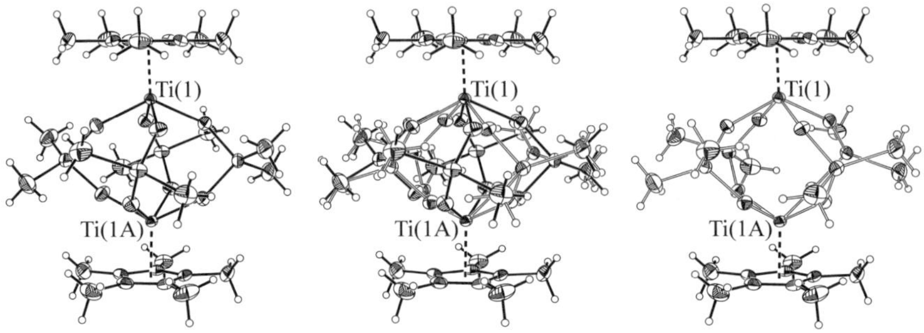

Fig. 5.16 Final model of $(\eta^5 C_5Me_5)_2Ti_2$ $(\mu\text{-}F)_8Al_4Me_8$. Left: component one, right: component two, middle: both components.

Comparing the first solution coming from SHELXS (Figure 5.9) to the final model, we can see that the solution (ti-00.res) already contained all non-hydrogen atoms. It was, however, not exactly easy to interpret that solution correctly.

5.3.3 A mixed crystal treated as occupancy disorder

The aluminum-imminosilicate $[RSi(NH)_3AlMeAlMe_2]_2$, where R is 2,5-$t Pr_2C_6H_3NSiMe_2 i Pr$, belongs to the same family as the molecule described in 5.3.1 (Rennekamp *et al.*, 2000). The only two differences are the metal (Al instead of Ga) and the alkyl groups attached to it (Me instead of Et). Stepwise halogenation of the aluminum with elementary iodine[11] leads to two different species, a twofold and a fourfold halogenated cage, whereby for each iodine atom added, one methyl group is eliminated. Both iodinated species crystallize together as a mixed crystal in the monoclinic space group $P2_1/n$ with half a molecule per asymmetry unit. The other half is generated from the first one by the crystallographic inversion centre. The model for this mixed crystal can be refined as disorder, assuming the presence of either a methyl group or an iodine atom at the same position. The usage of the `PART` instruction, the site occupancy factor and the second free variable is the same as described for positional disorder. In contrast to the 'normal case', where both components are identical with respect to atom types and number of atoms, while the coordinates of equivalent atoms are different, this mixed-crystal disorder is characterized by the opposite: `PART 1` and `PART 2` refer to different atom types, sharing the same coordinates. Even though the constraint `EXYZ`, which forces two atoms to occupy the exact same site, can be helpful for many occupational disorders, its use is not appropriate in this case, as the Al–C bond is significantly shorter than the Al–I bond.

[11] The same reaction also takes place with elementary bromine, giving rise to two- and fourfold brominated cages, which are entirely isostructural.

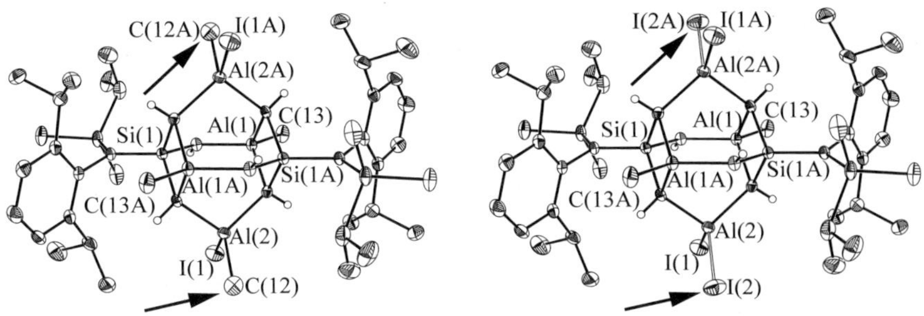

Fig. 5.17 Final model of the two differently iodinated aluminum-iminosilicates. On the lefthand side is the twofold iodinated cage, on the righthand side the fourfold iodinated one. Arrows point to the atoms that are different in the two molecules.

The second free variable refined to 0.85. This corresponds to 85% twofold iodinated product and 15% fourfold iodinated product. These numbers do not necessarily reflect the true ratio of the two products in the solution (and hence of the chemical reaction), as the halogenation significantly reduced the solubility of the cages. Thus, it can be theorized that the fourfold halogenated cage is even less soluble than the twofold halogenated species, which would lead to an overrepresentation of the fourfold iodinated molecule in the mixed crystal. Figure 5.17 shows the two different Al-Iminosilicates, and the files ali.res and ali.hkl on the accompanying CD-ROM contain the final model and the data, in case anyone wants to play around with this structure.

5.3.4 Disorder of solvent molecules

Solvent molecules, especially when they arc not coordinated to metal atoms, are quite often involved in disorders. They fill voids in the crystal lattice and can appear in several different orientations in the same void of different unit cells of a crystal, if these orientations are energetically approximately equivalent. Solvent molecules on special positions that do not possess the appropriate symmetry for these positions are also relatively common. In the spatial average this leads to disorder. As a matter of experience, some solvents are rarely involved in disorders (e.g. acetonitrile, which is linear and therefore has less opportunity to be disordered), while other solvents like chloroform seem to be disordered in almost every case.

Thetrahydrofuran (*thf*)
Tetrahydrofuran is a coordinating solvent and is, indeed, often found in crystal structures as an electronically neutral ligand coordinated to metal atoms. In such cases the oxygen atom is seldom disordered. However, rotation about the M–O axis is still possible (and often observed) and the ring itself, possessing an envelope conformation, may be disordered in several different ways. In the paragraph about

the restraint command SAME we already discussed an example of a completely disordered *thf* molecule, therefore I only show some pictures of typical *thf* disorders in this section (Figure 5.18).

Chloroform

One is more likely to find a disordered chloroform molecule in a crystal structure than a well-ordered one. Disorder of chlorine-containing molecules has stronger (negative) influence on the quality of the model than disorder of other solvents, because of the relatively high number of electrons in chlorine, which are disordered with the solvent. Therefore, whenever possible, one should avoid using chloroform or dichloromethane as crystallization solvent.

The refinement of chloroform is usually not difficult, even though it can be time consuming in some cases. Therefore it is not necessary to explicitly discuss an example here. Figure 5.19 shows a typical example of a disordered chloroform molecule.

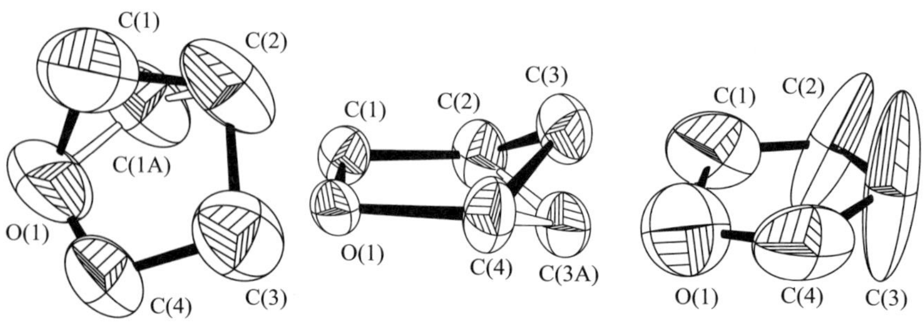

Fig. 5.18 Two typical tetrahydrofuran disorders (left and middle) and one example where disorder refinement of atoms C(2) and C(3) could improve the model.

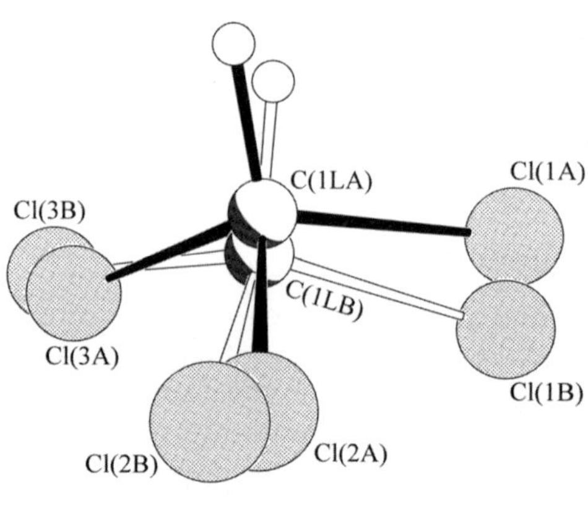

Fig. 5.19 Typical chloroform disorder. This $CHCl_3$ was taken from the last example structure of this chapter.

Toluene

Because of the methyl group, toluene (point group C_{2v} or, when you take into account the hydrogen atoms, even only C_s) possesses a much lower symmetry than benzene (point group D_{6h}). Nevertheless, toluene frequently occupies special positions in crystals compatible with the point group of benzene but not with that of toluene, for example inversion centres or twofold axes perpendicular to the aromatic ring plane. This behaviour appears as disorder in the spatial average. Besides that, toluene is often disordered even when it is not near a special position. Among many different possible disorders, there are two particularly frequently observed cases of toluene disorder.

The first case (see Figure 5.20) shows two discrete positions, the second position twisted relative to the first one by about 180°. Thus, the methyl group of one component lies close to the carbon atom C4 of the other component, in a way that both components are more or less coplanar one to another. An example for such a case will be given below.

The second case is characterized by a particularly undetectable methyl group.[12] The cause is a virtual rotation about the sixfold axis perpendicular to the molecule plane, which is part of the benzene molecule but not of toluene. Thus, in the spatial average the methyl group is distributed amongst six sites, contributing about one electron per position, located close to hydrogen atoms of other orientations. This disorder is best refined by ignoring it: such a toluene molecule should be refined as benzene.

The file tol-01.res on the accompanying CD-ROM contains a complete anisotropic model of a zirconium compound (for details see: Bai *et al.*, 2000) with hydrogen atoms but not the solvent yet. The space group is $C2/c$. Seven of the highest residual electron density peaks, Q(1) to Q(5), Q(8) and Q(9) (6.52, 6.10, 5.13, 5.10,

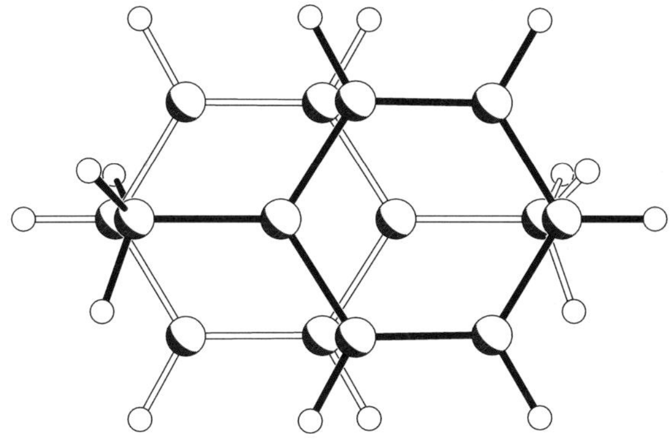

Fig. 5.20 A typical toluene disorder.

[12] The file ali.res, the third example in this chapter, is such a case.

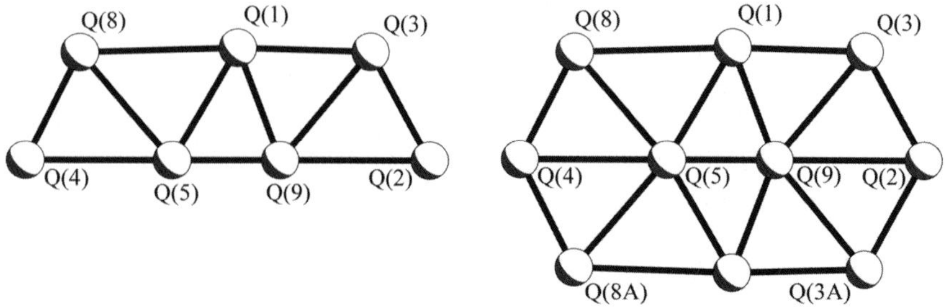

Fig. 5.21 Highest residual electron density maxima in tol-01.res forming a disordered toluene molecule on a mirror. Left-hand side: asymmetric unit; right-hand side: with symmetry equivalent atoms, revealing interpenetration toluene orientations.

4.93, 2.10, and 1.82 electrons per Å^3, respectively) are reasonably well separated from the main molecule, and seem to be some sort of solvent (see Figure 5.21A). From the crystallization conditions we suspect the presence of toluene, and the peaks Q(2), Q(4), Q(5), and Q(7) lie on the crystallographic mirror. After generating the symmetry equivalent atoms (e.g. using the GROW command in XP), one sees the shape of two interpenetrating toluene molecules, disordered as described above (Figure 5.21B). The following electron density maxima give rise to these carbon atoms:

Q(1) ➔ C(2A) and C(2B)
Q(2) ➔ C(4A) and C(10B)
Q(3) ➔ C(3A)
Q(4) ➔ C(10A) and C(4B)
Q(5) ➔ C(1A)
Q(8) ➔ C(3B)
Q(9) ➔ C(1B)

Thereby A belongs to `PART 1` (component 1) and B to `PART 2` (component 2).

Do not forget to change the site occupancy factors and to set the second free variable, as well as to give the similarity restraints (`SAME`) and `SIMU`, `DELU` and `FLAT` for the disordered atoms as it has been done in the file tol-02.ins. Take also into account the symmetry equivalents (use `EQIV`).

SHELXL produces the files tol-02.res and tol-02.lst. The second free variable refines to 0.75, which is a reasonable value, and the *R*-values have improved significantly. Figure 5.22 shows the two components in two different orientations. For the next step we can allow the disordered atoms to be refined anisotropically by adding `ANIS` to the next .ins file (tol-03.ins). Finally, we can add hydrogen atoms (tol-04.ins):

```
HFIX 43 C2A C2B C3A C3B C4A C4B
HFIX 123 C10A C10B
```

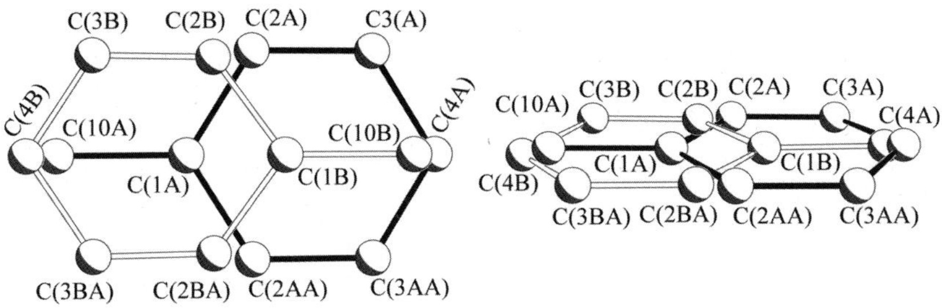

Fig. 5.22 First refined co-ordinates for the two toluene positions in tol-02.res.

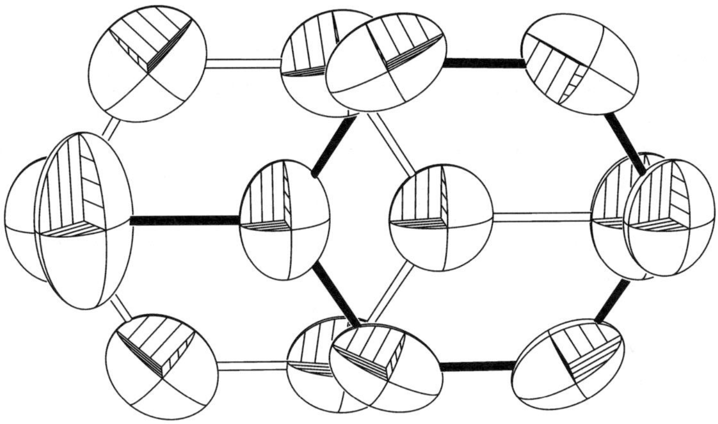

Fig. 5.23 Final model for the disordered toluene (hydrogen atoms omitted for clarity).

The refinement seems to be stable (results in tol-04.res and tol-04.lst), Figure 5.23 shows the final model. All that is left to do is to refine the weighting scheme to convergence, as it has been done in the file tol-05.res, which contains the publishable model.

Benzoic acid on a twofold axis

The file benz-01.res on the accompanying CD-ROM contains a complete anisotropic model of a diabetes drug,[13] which crystallizes in the monoclinic space group *C*2 with two molecules in the asymmetric unit. The model contains hydrogen atoms but not yet the solvent. The two independent molecules are related by a pseudo inversion centre, only violated by the chiral carbon atom. This one atom per molecule

[13] I am grateful to Alexander Pautsch and Herbert Nar of Boehringer Ingelheim Pharma GmbH & Co for providing the dataset and allowing me to use this structure as an example.

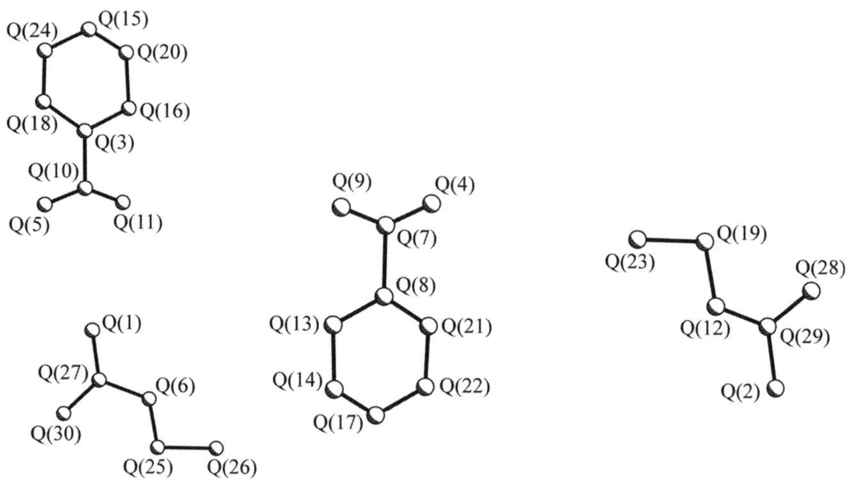

Fig. 5.24 The 30 highest residual electron density maxima in benz-01.res.

makes the difference between space group *C*2/m with one molecule per asymmetric unit and space group *C*2 with two independent molecules; but this point is not part of the disorder and will be the subject of Chapter 6. The 30 highest residual electron density peaks in benz-01.res, Q(1) to Q(30), with intensities of 2.24 to 2.17 electrons per Å^3, are significantly stronger than the others and seem to be solvent (see Figure 5.24). From the crystallization conditions we suspect the presence of benzoic acid or benzoate, and, indeed, the shapes formed by the peaks support this hypothesis. Two molecules of benzoic acid (or benzoate) are easily identifiable and—when assuming the atomic numbering scheme as shown in Figure 5.25—we can make the following assignments:

Q(3) ➔ C(11)
Q(5) ➔ O(12)
Q(10) ➔ C(17)
Q(11) ➔ O(11)
Q(15) ➔ C(14)
Q(16) ➔ C(12)
Q(18) ➔ C(16)
Q(20) ➔ C(13)
Q(24) ➔ C(15)

And

Q(4) ➔ O(21)
Q(7) ➔ C(27)

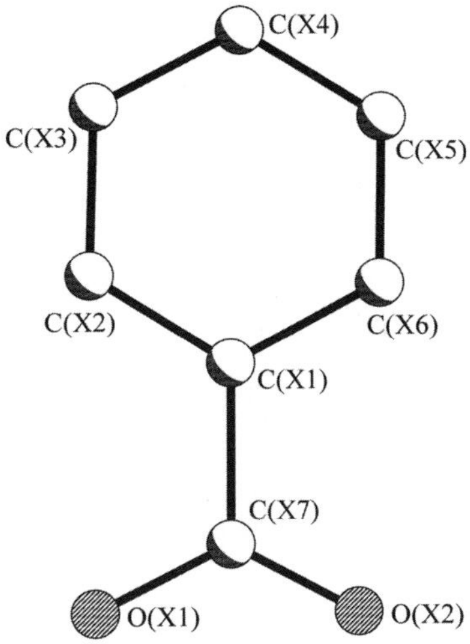

Fig. 5.25 Suggestion for an atomic numbering scheme for benzoate. The X stays for the molecule number (1 for the first benzoate molecule, 2 for the second, and so forth).

Q(8) → C(21)
Q(9) → O(22)
Q(13) → C(26)
Q(14) → C(25)
Q(17) → C(24)
Q(21) → C(22)
Q(22) → C(23)

Two more molecules are visible, though only partially. Taking into account the symmetry equivalents of the residual density peaks (e.g. using the `GROW` command in XP), one sees that these two molecules are located very close to crystallographic twofold axes. Even though benzoate possesses a twofold axis, in this case the molecules are not oriented along the crystallographic twofold axis but slightly tilted: the molecules are disordered (see Figure 5.26). As only the carboxyl group and three atoms of the aromatic ring are visible, we need to find a way to generate the missing atoms. The easiest is to use geometrical constraints as described before: `AFIX 66` generates a perfect hexagon. The following residual density maxima (see Figure 5.24) can be assigned.

Q(1) → O(32)
Q(6) → C(31)
Q(25) → C(32)
Q(26) → O(33)
Q(27) → C(37)
Q(30) → O(31)

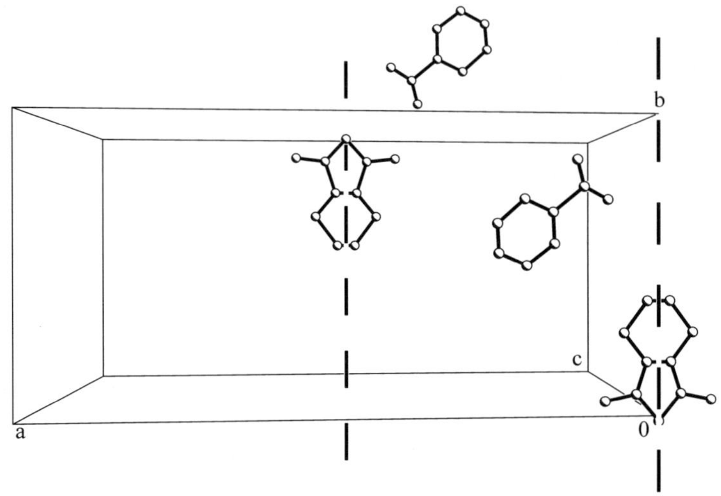

Fig. 5.26 Projection along the crystallographic *c*-axis (origin and unit cell axes are labelled) of the thirty highest residual density maxima. Two of the benzoate molecules are close to the crystallographic twofold axes (drawn as dashed vertical lines) and appear to be disordered.

And

Q(2) → O(42)
Q(12) → C(41)
Q(19) → C(42)
Q(23) → C(43)
Q(28) → O(41)
Q(29) → C(47)

The remaining three atoms of each ring are generated geometrically in the following way. Write `AFIX 66` in front of the first atoms of the phenyl rings (C(31) and C(41) respectively), complete the number of atoms (three atoms are missing in each of the two incomplete molecules) with atoms with the coordinates `0 0 0`, and write `AFIX 0` after the last atom of the two phenyl ring. Use `SAME` to restrain equivalent distances within and among the four solvent molecules. The site occupancy factor instructions for the two disordered molecules must be changed to `10.5000` and all disordered atoms must be in `PART -1`. As the second site of each disordered atom can be calculated directly from the positions of the atoms of the first component via the symmetry operator of the respective twofold axis, it is not necessary to have two parts in the .ins file. The negative part number suppresses the generation of special position constraints, and bonds to symmetry-related atoms are excluded from the connectivity table. In addition, the use of the second free variable is not indicated, as the ratio between the components is determined by the multiplicity of the special position, which is expressed by the *sof* instruction `10.5000`. `SIMU` and DELU have already been given for the entire structure earlier in the refinement, so we do not need to do it here for the disordered atoms. It is also important to add an AFIX 0 after the last hydrogen atom. This hydrogen atom was the last line before the HKLF 4 card, therefore an AFIX 0 would have been meaningless and SHELXL automatically removed it. If, however, other atoms follow, which is now the case, the HFIX 0 becomes important. Taking all this into account, the solvent part of the next .ins file (benz-02.ins) should look as follows (take some time to find out the meaning of the SAME commands):

```
AFIX 0
O11    4   0.36900    1.09370    0.50540   11.00000   0.05000
O12    4   0.41820    1.22620    0.47870   11.00000   0.05000
SAME C17 C11 C16 < C12
C17    1   0.38020    1.18030    0.49350   11.00000   0.05000
C11    1   0.33280    1.26340    0.49820   11.00000   0.05000
C12    1   0.28610    1.22030    0.51360   11.00000   0.05000
C13    1   0.24140    1.30520    0.51520   11.00000   0.05000
C14    1   0.24620    1.40130    0.50460   11.00000   0.05000
C15    1   0.29080    1.44490    0.48790   11.00000   0.05000
C16    1   0.33400    1.37520    0.48590   11.00000   0.05000
SAME O11 > C16
O21    4   0.07960    0.71580    0.01790   11.00000   0.05000
O22    4   0.13020    0.84580   -0.00650   11.00000   0.05000
C27    1   0.11870    0.75770    0.00450   11.00000   0.05000
C21    1   0.16800    0.67320    0.00210   11.00000   0.05000
C22    1   0.16490    0.56930    0.01220   11.00000   0.05000
C23    1   0.20910    0.49700    0.01270   11.00000   0.05000
C24    1   0.25450    0.53820   -0.00170   11.00000   0.05000
C25    1   0.25730    0.63580   -0.01390   11.00000   0.05000
C26    1   0.21400    0.71580   -0.01220   11.00000   0.05000
SAME O11 > C16
PART -1
O31    4   0.58950    0.88470    0.48860   10.50000   0.05000
O32    4   0.50000    0.95790    0.50000   10.50000   0.05000
C37    1   0.53770    0.87270    0.49580   10.50000   0.05000
AFIX 66
C31    1   0.51920    0.75480    0.49610   10.50000   0.05000
C32    1   0.55470    0.66620    0.49100   10.50000   0.05000
C33    1   0.51570    0.55780    0.49670   10.50000   0.05000
C34    1   0          0          0          10.5       0.05
C35    1   0          0          0          10.5       0.05
C36    1   0          0          0          10.5       0.05
AFIX 0
SAME O11 > C16
```

```
O41    4 -0.08800   0.05130   0.00970  10.50000  0.05000
O42    4  0.00000  -0.01740   0.00000  10.50000  0.05000
C47    1 -0.03700   0.07310   0.00530  10.50000  0.05000
AFIX 66
C41    1 -0.01990   0.17860   0.00210  10.50000  0.05000
C42    1 -0.05710   0.27550   0.00820  10.50000  0.05000
C43    1 -0.01870   0.38040   0.00340  10.50000  0.05000
C44    1  0         0         0        10.5      0.05
C45    1  0         0         0        10.5      0.05
C46    1  0         0         0        10.5      0.05
AFIX 0
PART 0
```

The file benz-02.res contains the complete disorder. Now the `AFIX 66` and `AFIX 0` lines can be removed,[14] and the atoms can be refined anisotropically (add `ANIS`), as the disorder is stable. This has been done in the file benz-03.ins.

In the peak list of residual density maxima found in the file benz-03.res, the hydrogen atoms on the non-disordered phenyl rings appear quite clearly. `HFIX 43` applied to these positions in the next step (benz-04.ins) generates the hydrogen atoms on geometrically calculated positions. In the file benz-04.res the refinement of the disorder is complete. Before publishing the structure, however, there are a couple of questions to be addressed. First: how many benzoic acid (or benzoate) molecules are there per molecule of the drug? The answer is one-and-a-half.[15] There are clearly two complete independent molecules of the drug and two fully occupied and not disordered solvent molecules per asymmetric unit. Also, there are the two disordered half molecules in the asymmetric unit. In the complete unit cell, there are four drug molecules and six solvent molecules, two of the latter located on special positions in a disordered way. The second question is: are the solvent molecules benzoic acids or benzoates? To answer this, the total charge of the rest of the asymmetric unit needs to be taken into account. The drug molecules each bear one positive charge on the nitrogen atom N(1) (the three hydrogen atoms were clearly visible in the difference Fourier for both independent molecules), which makes it necessary that two of the three solvent molecules in the asymmetric unit be benzoate ions and the third one benzoic acid. Looking in the residual electron density for the one missing hydrogen atom (it might actually be disordered over the eight possible positions) and thinking about possible hydrogen-bonding patterns can be a nice weekend-pastime for the inclined reader. Taking the two strongest residual density maxima, Q(1) and Q(2) each as half a hydrogen atom leads to a scenario in which both the disordered benzoic acid molecules connect two (symmetry equivalent) benzoate ions in the unit cell, as shown in Figure 5.27 for one of the two independent benzoate ions.

[14] You will find that the last `AFIX 0` has automatically been removed by SHELXL, as it was in the last line before the `HKLF 4` card and hence meaningless. So there are only three lines left to be removed.

[15] And not two, as many—even experienced—crystallographers might answer. Disordered molecules on special positions are a famous and infamous trap and it is sometimes hard to picture such a scenario correctly.

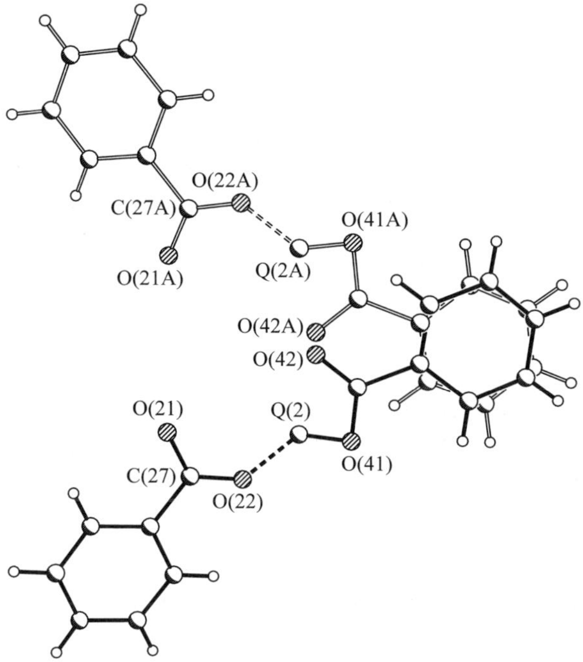

Fig. 5.27 Possible hydrogen bonding pattern for one of the two independent benzoate molecules. Atoms of symmetry equivalent atoms are labeled with an A after the original atom name.

5.3.5 Three types of disorder in one structure: cycloikositetraphenylene

The cycloikositetraphenylene[16] shown in Figure 5.28 crystallizes in the rhombohedral space group $R\bar{3}$ with a sixth of the macrocycle and one molecule of chloroform in the asymmetric unit. The rest of the molecule and five more $CHCl_3$ molecules are generated by the $\bar{3}$ axis. The macrocycle consists of six units of four 1,4-linked benzene rings, bonded via 1,3 links to one another. Thus, a hexagon containing 24 benzene rings is formed. The edges of the hexagon consist of five benzene rings, the rings at the corners being shared by two edges. The middle ring of each edge carries two hexyl groups *trans* to one other. One of them is directed towards the centre of the macrocycle, and the other one points out of it. The structure analysis was a challenge and difficult in all stages, save the solution of the phase problem (Müller *et al.*, 2001). After many unsuccessful trials, crystals could only be obtained from chloroform. These crystals were very unstable when removed from

[16] In the publications describing the synthesis and structure of this molecule, it is called cyclotetraicosaphenylene, reflecting the name used on previous publications. Taking into account that a body with 24 sites is not called tetraikosahedron but rather ikositetrahedron, I prefer the name cycloikositetraphenylene for the molecule described on these pages.

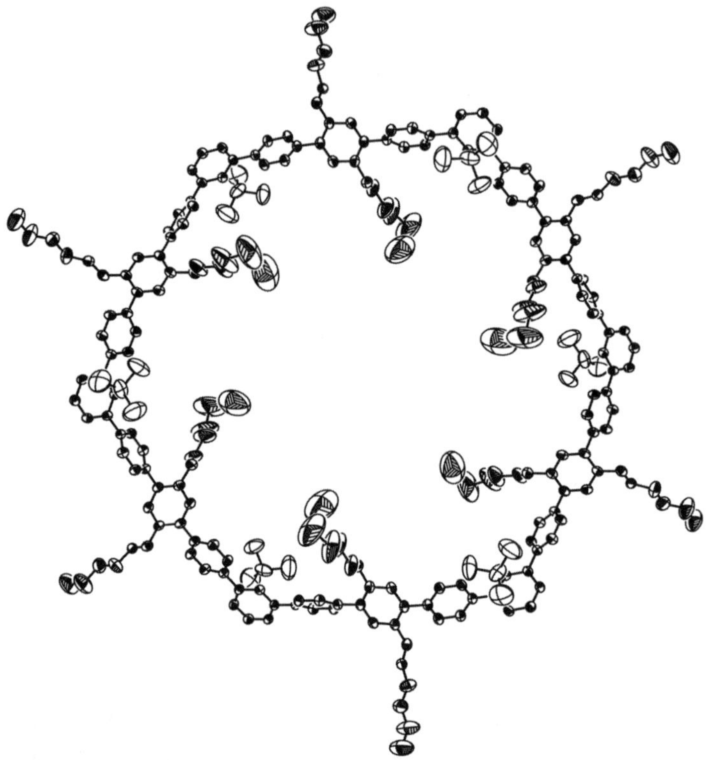

Fig. 5.28 Crystal structure of the Cycloikositetraphenylene showing 50% ellipsoids. The second components and hydrogen atoms have been omitted for clarity.

the mother liquor at room temperature (see Figure 5.29) and could only be mounted onto the diffractometer using cryo-techniques at all times. The diffraction pattern also showed some anomalies (see Figure 5.30): the low resolution data have a very high noise level, and the reflection profile appears to be strangely elongated in some orientations, which created several problems with the data reduction. The solution from SHELXS already contained all atoms of the aromatic skeletal structure. The positions of the *n*-hexyl chains, however, were missing. These hexyl chains turned out to be highly disordered and give a good example for continuous disorder. As the difference between discrete disorder and diffuse movement is fluid, and since refinement of more than two sites for the disordered hexyl groups was not stable, the model was reduced to the two main components for each hexyl group, accepting relatively large anisotropic displacement parameters. As shown in Figure 5.31, the disorder causes about half of the hexyl groups to lie above the ring plane, and the other half to lie beneath it (the site occupancy factors refined to 0.50/0.50 for the hexyl chains pointing out of the macrocycle, and to 0.55/0.45 for the other hexyl groups). Thus, a fork-like structure is formed, which seems to be favourable for the

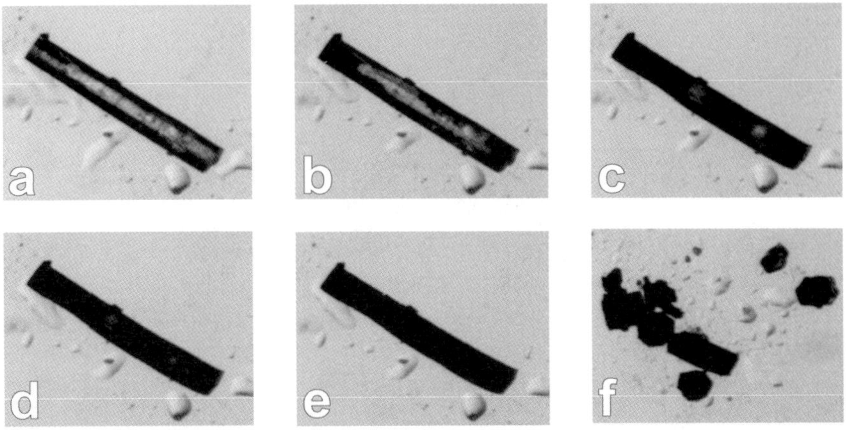

Fig. 5.29 Crystal of the cycloikositetraphenylene under the microscope directly after taking it from the flask (a), after 10 seconds (b), after 30 seconds (c), after 40 seconds (d), after 60 seconds (e) and after touching the crystal with a needle (f).

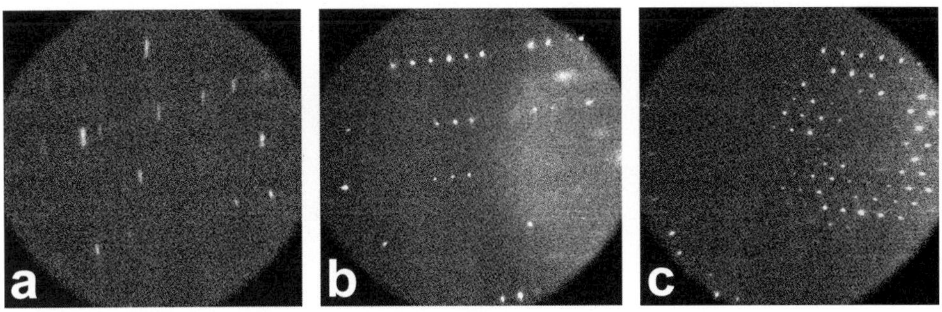

Fig. 5.30 Three diffraction pictures of the cycloikositetraphenylene. a: unusual reflection profile in some orientations, b: high background at low resolution, c: a 'good looking' frame.

crystal packing. In contrast, the positions of the benzene-ring atoms are well-defined, and their anisotropic displacement parameters are relatively small.

As mentioned above, the asymmetric unit also contains one molecule of chloroform. This molecule is linked to one of the phenyl rings via a weak CH–π hydrogen bond (see Figure 5.32). In addition, the $CHCl_3$ molecule is disordered approximately about the C–H$\cdots\pi$-axis. Figure 5.19 shows this disorder.

The whole macrocycle is not planar but, in accordance with the $\bar{3}$-geometry, possesses a cyclohexane-like chair conformation. In the three-dimensional packing the hexagons are stacked like coins, or more precisely like garden chairs. The disordered hexyl chains dovetail with the $CHCl_3$ molecules and with other hexyl groups. Although this part of the packing is very compact, the relatively large holes at the centre of the molecules lie directly over one another, giving rise to potentially

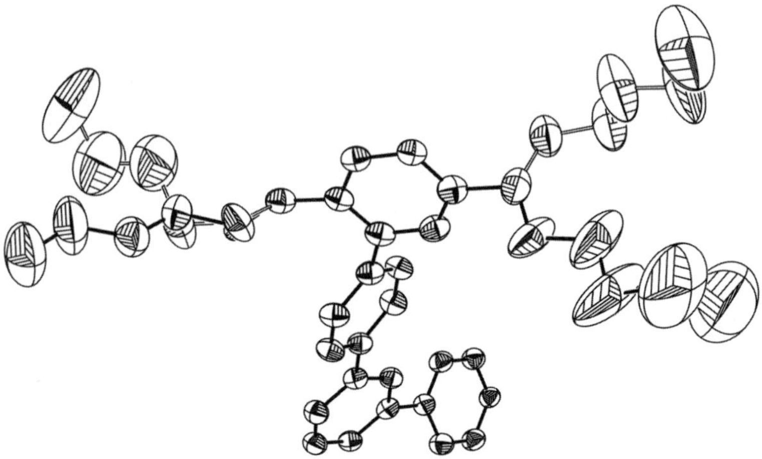

Fig. 5.31 Fork-like disorder of the *n*-hxyl groups in the structure of the cycloikositetraphenylene.

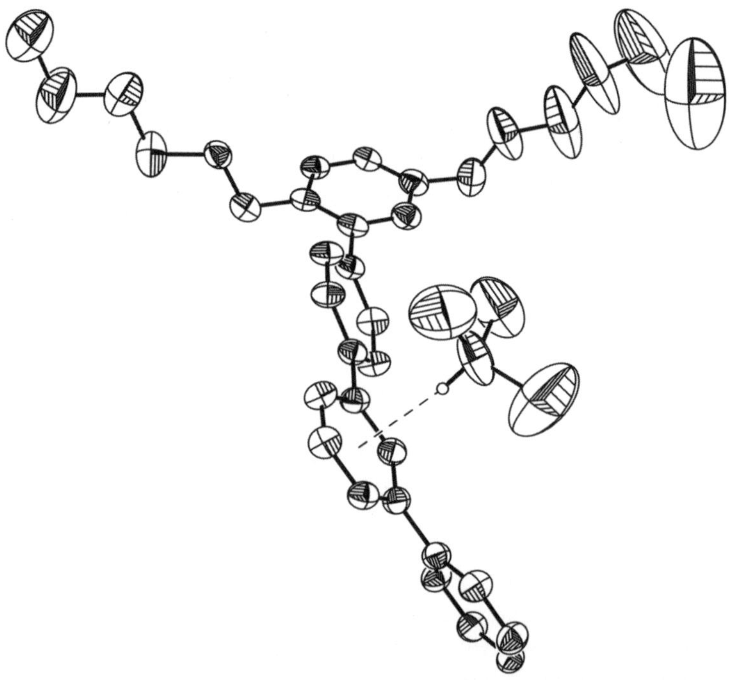

Fig. 5.32 CH-π-interaction between the chloroform molecule and a phenyl ring of the cycloikositetraphenylene. Distances and angles: C $\cdots$ π: 3.38 Å, H $\cdots$ π: 2.38 Å; C-H-π: 174.3°.

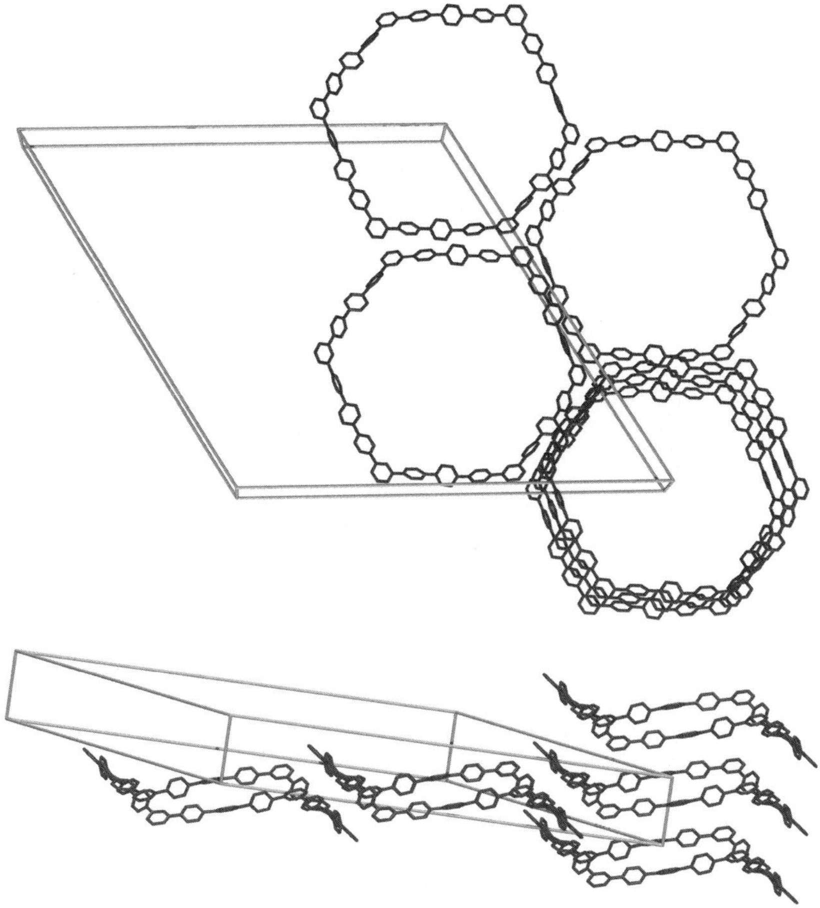

Fig. 5.33 View along the crystallographic c-axis (above) and side view (below) of the crystal packing of the cyclotetraikosiphenylene. Hydrogen atoms, chloroform molecules and hexyl chains have been omitted for clarity.

endless tubes through the whole crystal (Figure 5.33). The interior of these channels appears to be completely empty, as no solvent molecules can be located nor any relevant residual electron density can be found in the difference Fourier syntheses. Since the absence of any matter in cavities of this size is impossible, it is probable that they are filled with liquid $CHCl_3$ like the pores of a sponge. During data collection at 133 K, the $CHCl_3$ is amorphously frozen. This explains the behaviour of the crystals under the microscope: when removed from the mother liquor at room temperature, the crystals lose the $CHCl_3$ from the channels and the lattice breaks down. This hypothesis also helps to explain the appearance of the diffraction pattern: the unusual high background at very low resolution is a result of the diffuse scattering of the chaotically disordered $CHCl_3$ molecules; the high mosaicity of some reflections could be the result of small, randomly distributed damage in the crystal lattice caused by evaporation of some of the $CHCl_3$ molecules.

Consequently, a bulk solvent correction was performed; this significantly improved the *R*-values of the refinement and the standard deviation of the bond lengths and angles. Following Babinet's principle, SHELXL refines two parameters: the first grows with the amount of diffuse solvent, and usually possesses values around one when the mean electron densities of the solvent and ordered parts of the structure are similar, as in most protein crystals. A large value of the second parameter indicates that only the low-angle data are influenced by the diffuse scattering of the bulk solvent; values of 2 to 5 are typical. In the case of the liquid $CHCl_3$, the first parameter refined to 13, the second to 9, which indicates extremely large regions of disordered solvent of higher mean electron density than the rest of the structure, affecting exclusively the low-resolution data.

The file cyclo.hkl on the accompanying CD-ROM contains the dataset; the files cyclo-0.res and cyclo-x.res correspond to the solution from SHELXS and the final, publishable model. The interested reader may try to proceed from one to the other.

6

Pseudo-Symmetry

Many, maybe even most crystals contain one crystallographically independent molecule, but sometimes there is only half a molecule (or less) in the asymmetric unit, and a crystallographic symmetry element generates the remainder of the molecule. In fact, several of the examples in this book are structures of this type (for example Section 5.3.5, where the asymmetric unit contains only a sixth of the molecule, and all three examples in Chapter 4). The opposite and somewhat less common effect is to find more than one molecule in each asymmetric unit.

Having more than one molecule in the asymmetric unit occurs predominantly in space groups of low symmetry like $P\bar{1}$ or $P2_1$. In most of these cases the two (or more) independent molecules are not related by simple symmetry operators such as twofold axes, mirror planes or inversion centres, but are different rotamers of the same molecule. Those cases are not what this chapter is about. This chapter deals with structures where there is in fact non-crystallographic symmetry to be found, relating two or more crystallographically independent molecules.

There are two cases to be distinguished: true non-crystallographic symmetry (NCS), and global pseudo-symmetry. The former, NCS, has usually no negative effect on the refinement (other than taking more time picking and naming all the atoms from the difference Fourier map) and can be seen as a—sometimes even useful—curiosity. The latter, however, can cause systematic errors which need to be addressed by the crystallographer. In the example section of this chapter we will discuss one of each. The first example is a relatively tricky case of global pseudo-symmetry, in which there is a choice between space groups Pn and $P2_1/n$. This example also illustrates that pseudo-symmetry does not need several molecules in the asymmetric unit—the pseudo-symmetry operator can be located within the molecule itself. The second example is a case of multiple non-crystallographic symmetry.

Generally, it is important to check very carefully whether an example of pseudo-symmetry is not, in fact, a case of overlooked crystallographic symmetry, and hence a case of wrong space group. Many published structures were assigned incorrect space groups, almost all of them space groups with too low symmetry, but this topic is beyond the scope of this chapter and will be addressed in Chapter 9.

6.1 Global pseudo-symmetry

In the case of global pseudo-symmetry, two molecules in the asymmetric unit[1] are *almost but not quite* related by a crystallographic symmetry operator of a higher-symmetric space group. Crystallographic symmetry elements correspond to special positions, which make them valid throughout the entire crystal. If, for example, in space group $P2_1$ two independent molecules are oriented in a way that they are almost but not quite related by a glide plane along c and perpendicular to the monoclinic axis, we have a case of global pseudo-symmetry, and the pseudo space-group is $P2_1/c$. As the glide plane is almost but not quite fulfilled, the associated zonal systematic absences are almost but not quite absent. That means reflections of the type $h\ 0\ l$ with $l \neq 2n$ are very weak but most of them still significantly present. This effect can add additional ambivalence to space group determination. The biggest problem, however, is that global pseudo-symmetry creates systematic errors. The fact that non-equivalent atoms are almost but not quite related by a crystallographic symmetry operator leads to sometimes strong correlation between two positions (check the list of 'largest correlation matrix elements' in the .lst file). This correlation can lead to geometrical distortion (deviations in bond-distances and angles) or problems with anisotropic refinement and can be resolved by the use of restraints and/or constraints. In some cases, when the violation of symmetry is only marginal, it can be appropriate to choose the higher-symmetry space group and refine a disorder; in other cases refinement in the lower-symmetry space group is better. The first example in this chapter deals with such a case.

6.2 True NCS

'Non-crystallographic symmetry' or NCS means that two (or more) crystallographically independent molecules are perfectly or almost perfectly related by a symmetry element like an inversion centre or a rotation axis that is not part of the space group symmetry. Not being part of the space group symmetry means that this symmetry element is not on a special position. Such symmetry is valid only within one single unit cell and not—like real crystallographic symmetry or global pseudo-symmetry—throughout the crystal. It is important to make sure that a non-crystallographic symmetry element cannot be transformed into a crystallographic one by simply re-arranging the unit cell setting (e.g. switching or halving unit cell axes). If such a different cell setting can be found, true NCS can be transformed into global pseudo-symmetry or, if the two molecules overlap perfectly, there may even be no pseudo-symmetry at all in the new setting.

True NCS occurs much more frequently than global pseudo-symmetry. The only problem it causes is the somewhat longer time and effort it takes to refine the structure, simply due to the much larger number of independent atoms. On the other hand, in some cases—for example for a protein structure with a very low data-to-parameter

[1] Of course, global pseudo-symmetry is, in principle, also possible for more than two molecules (say six molecules almost but not quite related by a crystallographic sixfold) or for only a fraction of a molecule (as in the first example of this chapter).

ratio (say at 3.5 Å)—non-crystallographic symmetry can even be helpful: assuming the space group does not change, the number of observable reflections to a certain resolution increases with the size of the unit cell and so usually does the number of parameters that need to be refined.[2] However if the volume of a unit cell is larger because of non-crystallographic symmetry, distances and angles of the NCS-related molecules can be associated with the help of restraints (`SAME`, `SADI`), which indirectly increases the amount of data, thus improving the data-to-parameter ratio. The second example in this chapter describes a structure with six independent molecules in the asymmetric unit, related by two pseudo-twofold axes and a pseudo-inversion centre.

6.3 Examples

In the following sections I present two examples of pseudo-symmetry. All files you may need in order to perform the refinements yourself are given on the CD-ROM that accompanies this book. The first case describes a molecule that is located on a crystallographic inversion centre without actually fulfilling the symmetry. This arrangement results in global pseudo-symmetry. The second example is a case of true non-crystallographic symmetry with six crystallographically independent molecules.

6.3.1 Pn or $P2_1/n$

The crystal structures of two very similar molecules from the large family of indigoid dyes were determined. The refinement of one structure was straightforward and did not pose any difficulties, while the other structure caused some headaches. The two molecules have the impressive names *trans*-5-(4,4-dimethyl-3-oxo-thiolan-2-ylidene)-3,3-dimethyl-[1,2′]dithiolan-4-one and trans-5,5,5′,5′-tetramethyl-[3,3′]bi[1,2′]di-thiolanylidene-4,4′-dione, and for the rest of this chapter I will not mention their names again but rather use the numbers **1** for the first molecule and **2** for the second one. Figure 6.1 shows the two molecules side by side, and if not from their names, then from the figure, it can be seen easily that the two molecules differ only by the substitution of an S atom with a CH_2 group. This reduces the symmetry of **1** with respect to **2**: molecule **2** (point group C_{2i}) has an inversion centre in the middle of the C–C bond between the two five-membered rings, while molecule **1** (point group C_2) does not. However, the deviation of **1** from point group C_{2i} is only marginal, as only one non-hydrogen atom breaks the symmetry. The chemistry behind these two molecules is described in Gerke *et al.* (1999).

Molecule **2** crystallizes in the monoclinic space group $P2_1/n$ (unit cell dimensions: $a = 8.265(2)$, $b = 8.228(2)$, $c = 9.178(2)$, $\beta = 101.14(3)$) with half a molecule in the asymmetric unit. The other half is generated by the inversion centre of the

[2] The number of parameters depends mostly on the number of atoms in the asymmetric unit. That means if a unit cell is larger owing to higher bulk-solvent content, only the number of observable reflections would change and not the number of parameters to be refined.

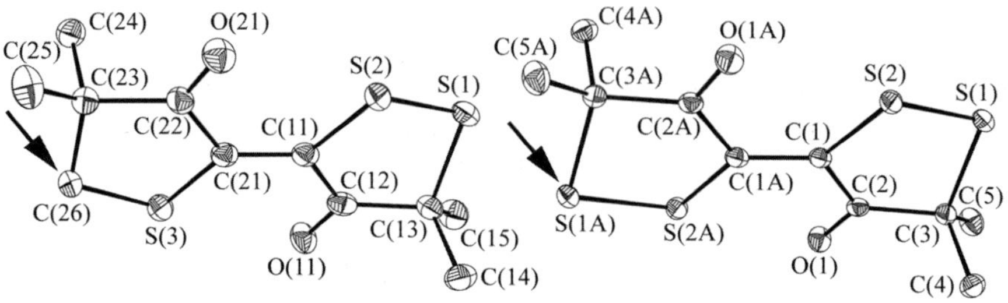

Fig. 6.1 Indigo derivatives **1** (left) and **2** (right). Sulfur atom S(1A) in **2** corresponds to carbon atom C(26) in **1** (see black arrows); this is the only difference between the two molecules.

Table 6.1 Systematic absences statistics for the dataset of **1**

	-21-	-a-	-c-	-n-
N	33	609	618	585
N I>3s	17	311	312	1
<I>	5.9	61.3	60.4	0.2
<I/s>	9.1	12.9	12.7	0.5

space group. The unit cell of **1** is almost identical: $a = 8.4162(1)$, $b = 8.1001(9)$, $c = 9.209(1)$, $\beta = 100.953(7)$. However, as the molecule lacks an inversion centre, it cannot crystallize in the same space group without problems, unless the complete molecule dwells in a discrete positional 1:1 disorder about the crystallographic inversion centre. This case can be described as a disorder of only those atoms in violation of the C_{2i} symmetry, that is the CH_2 group and the third S atom. The file s-00.hkl on the accompanying CD-ROM corresponds to the absorption-corrected, but unmerged diffraction data for this molecule. If you analyze this data, for example using a program like XPREP, you will get the following systematic absences statistics as shown in Table 6.1.

In Table 6.1, the four rows tell us the following: `N` is the number of independent reflections that should be absent if the respective symmetry element is present. The second row describes how many of those `N` reflections are stronger than three times their own standard uncertainty. `<I>` in the third row is the average intensity of the `N` reflections and `<I/s>` in the last row describes the average I/σ value of the reflections that should be absent in the presence of the respective symmetry element. The systematic absences clearly indicate the presence of a glide plane in the direction of n and the absence of an a or c glide plane. The situation for the twofold screw axis is less clear. Half of the 33 reflections $0\ k\ 0$ with $k \neq 2n$ that should be absent for a monoclinic 2_1-axis are observed, however those observed are significantly weaker than the rest of the data.

The $|E^2 - 1|$ statistic is near the expected value for a centrosymmetric crystal ($|E^2 - 1| = 0.885$), which suggests space group $P2/n$ or $P2_1/n$ rather than Pn.[3] Taking into account our knowledge about the size and geometry of the molecule, these two space groups will work only if the third sulfur atom is disordered against the CH_2 group (as described above). The program XPREP suggests $P2/n$, and the files s01.ins and s-01.hkl are the SHELXS input files set up and merged in this space group.

SHELXS seems to find a solution, but a closer look at the suggested atomic positions in the file s-01.res quickly shows that this is only a pseudo-solution. Other attempts like trying other seminvariants,[4] Patterson methods or using the program SHELXD (Usón and Sheldrick, 1999)[5] do not result in a correct solution in this space group either. Therefore we need to try our luck in other space groups: the files s-02.ins and s-02.hkl are the SHELXS input files set up and merged in space group $P2_1/n$.

The $P2_1/n$ solution from SHELXS is in the file s-02.res. When looking at the solution with a graphical interface like Ortep or XP, you see the following: one sulfur atom has been correctly identified and the electron density maxima Q(1) to Q(7) correspond to the remaining non-hydrogen atoms. Q(1) is significantly higher than all the other peaks in the list and corresponds to the one missing sulfur atom. However, as mentioned above, we know that it can only be a half-occupied sulfur atom superimposed onto a half-occupied CH_2 group. Using our disorder-refinement skills from Chapter 5, we can formulate this as a disorder. Note that this disorder is about a crystallographic inversion centre and hence the ratio of the two components is necessarily 1:1. This makes the use of a second free variable unnecessary; we can simply set the occupancies of the affected atoms to 0.5. All this has been done in the SHELXL input file s-03.ins. The results after 10 cycles of least squares refinement (s-03.res) appear promising. We can try to refine the molecule anisotropically (s-04.ins).

Most of the anisotropic model looks reasonable, and all hydrogen atoms can be found in the difference Fourier map. However, the disordered carbon, atom C(6), is non-positive definite (NPD). A quick temporary fix is to constrain the anisotropic displacement parameters of C(6) to be identical with those of S(2). This can be

[3] The $|E^2 - 1|$ statistic is explained in detail in Chapter 7. In short, the value for $|E^2 - 1|$ reflects the relative intensity distribution in reciprocal space. Lower values correspond to more evenly distributed intensities as they are found for non-centrosymmetric space groups, while higher values point to more intensity fluctuations in the diffraction pattern, which is typical for centrosymmetric space groups. There are two expected values for $|E^2 - 1|$: 0.736 for non-centrosymmetric structures and 0.968 for centrosymmetric ones. The value of 0.885 in the example is closer to 0.968 than 0.736, which lets us expect a centrosymmetric space group like $P2/n$ or $P2_1/n$ rather than the non-centrosymmetric space group Pn.

[4] Sometimes SHELXS finds two groups of potential solutions, where one corresponds only to a pseudo-solution. If this pseudo-solution has the lower combined figure of merit of the two, the program will chose it over the correct solution. In such a case it is common practice to examine the SHELXS .lst file, identify the other possible solution and run the program a second time specifying the seminvariants in the `TREF` command. This book is about structure refinement, not the solution of the phase problem and for further details the reader is referred to the SHELX manual or the original publication (Sheldrick, 1990).

[5] Even though SHELXD was written specifically for macromolecular problems, this program can also be used quite successfully for the solution of small molecule structures. The latest version of the program even supports the `TWIN` command (see Chapter 7), which makes it particularly useful for the solution of twinned structures of any size.

achieved by adding the line `EADP C6 S2` to the next .ins file, s-04.ins. This file also contains the `HFIX` commands for all hydrogen atoms (see Chapter 3).

The model in s-04.res looks good, and SHELXL suggests an extinction correction. Extinction is the weakening of a reflection owing to secondary diffraction. This effect is relatively weak and requires large crystals of very high quality to become noticeable. Indeed, the crystal used in the diffraction experiment was large (0.4 × 0.4 × 0.3 mm) and very good.[6] Adding the command `EXTI` (Larson, 1970) in the header of the next .ins file, s-05.ins, should take care of this. We should also try to remove the `EADP` command and start adjusting the weighting scheme.

Whenever extinction is refined, it is important to check in the .lst file whether the extinction coefficient refines to sensible values with a reasonably small standard uncertainty. This is the case here, so we will keep `EXTI` in. Unfortunately, removing the `EADP` constraint causes the carbon atom C(6) to be NPD again. This is not unusual for global pseudo-symmetry, as the abovementioned correlation among symmetry-related but not equivalent atoms can lead to problems with anisotropic refinement. Therefore, we decide to live with this constraint and finalize the refinement by adjusting the weighting scheme. The final model in $P2_1/n$ is in the file s-05a.res.

Although satisfactory, this model is by no means perfect: the final residual values are very high ($R1 = 0.0644$ for $F_o > 4\sigma F_o$ and $wR2 = 0.1576$ for all reflections)[7] when compared with the much lower merging R-values ($R_{\text{int}} = 0.0287$, $R_{\text{sigma}} = 0.0116$),[8] and the first coefficient of the weighting scheme alarmingly refines to zero. We should also try the third possible space group, *Pn*. That means starting over with the file s-00.hkl. The files s-06.ins and s-06.hkl are the new SHELXS input files, set up and merged in *Pn*.

The solution from SHELXS is in the file s-06.res. When looking at the solution with a graphical interface, you see the following: two sulfur atoms have been correctly identified and the electron density maxima Q(1) to Q(14) correspond to the remaining non-hydrogen atoms. Either Q(1) or Q(2) must correspond to the one missing sulfur atom. Both peaks Q(1) and Q(2) are significantly higher than all the other peaks in the list, and unfortunately, they are about equal. This makes the choice rather difficult and it is possible that they both correspond to the missing sulfur atom, if the position of the sulfur is disordered against the position of the CH_2 group in the same way as we assumed it for the centrosymmetric case. The only difference in *Pn* is that the disorder is no longer about a crystallographic symmetry element. This makes the use of a second free variable necessary, as the ratio between the two components can assume any value. The atom type assignment and the formulation of this disorder (with restraints) have been done in the file s-07.ins.

The model after 10 cycles of refinement with SHELXL (s-07.res) looks reasonable, even though the second free variable refined to a rather high value (0.92(1)) and the U_{eq} values for the disordered atoms are not very similar. Expecting from the fluctuating U_{eq} values that at least one of the disordered atoms will become

[6] In fact, SHELXL is not necessarily right when it suggests the performance of an extinction correction. Other effects can mimic extinction and SHELXL cannot easily distinguish between certain bulk solvent effects and extinction.

[7] A definition of the R-values is given in Chapter 2: Equations 2.3 and 2.4.

[8] The merging R-values are introduced in Chapter 1: Equations 1.1 and 1.2.

NPD, we use two EADP commands as described above and try to refine all atoms anisotropically (add ANIS in s-08.ins).

The anisotropic model in s-08.res looks good and all hydrogen atoms are clearly visible in the difference Fourier map. The file s-09.ins contains the HFIX commands for all hydrogen atoms.

Pn is a non-centrosymmetric space group. Therefore, in the presence of atoms heavy enough to cause anomalous scattering at the given wavelength,[9] we must check the Flack-*x* parameter (Flack, 1983), which can be found in the .lst file, directly after the final structure factor calculations. This parameter is supposed to be approximately zero for the correct absolute structure, approximately unity for the inverted absolute structure (within the standard uncertainty) and possesses values between one and zero for racemic twins. More details about this parameter can be found in Chapter 7 of this book. In the file s-09.lst, the Flack-*x* parameter is listed as 0.51(9). Therefore, we need to refine a racemic twin (see Chapter 7 for details). The file s-10.ins contains the lines TWIN and BASF 0.6 to accommodate the racemic twinning.[10] In addition, we can start adjusting the weighting scheme.

The final model with adjusted weighting scheme is given in the file s-11.res. It is not possible to release the two EADP constraints, but the final *R* values are much better than for the model in the centrosymmetric space group ($R1 = 0.0265$ for $F_o > 4\sigma F_o$ and $wR2 = 0.0657$ for all reflections).[11] Clearly, space group *Pn* describes the structure better than $P2_1/n$, and the crystallographic inversion centre is the pseudo-symmetry operator in this case of global pseudo-symmetry. As described above, global pseudo-symmetry frequently results in strong systematic errors in the refinement, which, in this case, can only be overcome with the use of two ADP constraints. The fact that the structure shows almost but not quite a twofold screw axis along *b* explains the partially fulfilled systematic absences. The value for $|E^2 - 1|$ (0.885), which is, as discussed above, too high for a non-centrosymmetric space group, can be explained with the fact that the structure is indeed pseudo-centrosymmetric. This gives rise to a pseudo-centrosymmetric $|E^2 - 1|$ statistic.

In 1999, this structure was published in space group $P2_1/n$—yes, by the author of this chapter—and it was not until the preparations for this book, that the structure was revisited and that the better description in the lower-symmetry space group was made.

6.3.2 $[Si(NH_2)_2CH(SiMe_3)_2]_2$: $P\bar{1}$ with $Z = 12$

The compound $[Si(NH_2)_2CH(SiMe_3)_2]_2$ crystallizes in the triclinic space group $P\bar{1}$ with six independent molecules in the asymmetric unit. The molecules consist

[9] For Mo radiation anything heavier than Si qualifies. With Cu radiation sometimes even oxygen shows significant anomalous signal.

[10] A value of 0.5 for the Flack-x parameter points to a 50:50 twin, corresponding to a value of 0.5 for the BASF. Frequently, however, a starting value of 0.5 for free variables or batch scale factors corresponds to a pseudo-minimum. It is better to start with values slightly above or below 0.5, for example 0.4 or 0.6.

[11] You may have noticed that the refinement in *Pn* does not contain the EXTI command. This is because attempts to refine extinction did not give rise to a significant improvement of the model (see s-12.res).

of a $(NH_2)_2Si–Si(NH_2)_2$ core, where the tetrahedral coordination sphere of each silicon atom is completed by a $CH(SiMe_3)_2$ ligand. Four of the molecules show syn-periplanar conformation, two possess the anti-clinal conformation. A detailed description of the related chemistry can be found in Ackerhans *et al.* (2001).

The files sin-01.ins and sin-01.hkl on the accompanying CD-ROM are the SHELXS input files set up and merged in $P\bar{1}$. The solution contains all 144 non-hydrogen atoms: the atom type of the 36 silicon atoms have been assigned correctly, and the electron density peaks Q(1) to Q(107) and Q(111) correspond to the carbon and nitrogen atoms. Assigning, naming and sorting all atoms takes a while and has been done in the file sin-02.ins. The first refined model in sin-02.res is stable and we can refine all atoms anisotropically right away. The file sin-03.ins contains the `ANIS` command, and the `PLAN` card has been changed to 300 in order to give enough residual density maxima to show all or most hydrogen positions. The file sin-03.res is the complete anisotropic model and most hydrogen positions can be found in the difference Fourier. The positions of hydrogen atoms bound to carbon can be calculated and refined using a riding model with the help of `HFIX` commands, which has been done in the file sin-04.ins.

In the file sin-04.res, the following residual density maxima correspond to hydrogen atoms bound to nitrogen: Q(17), Q(20), Q(45), Q(60), Q(32), Q(49), Q(56), Q(63) for molecule 1; Q(23), Q(37), Q(53), Q(67), Q(50), Q(66), Q(31), Q(51) for molecule 2; Q(24), Q(30), Q(23), Q(64), Q(9), Q(13), Q(28), Q(96) for molecule 3; Q(85), Q(207), Q(71), Q(94), Q(15), Q(52), Q(12), Q25) for molecule 4; Q(22), Q(27), Q(35), Q(36), Q(14), Q(47), Q(11), Q(38) for molecule 5; and Q(10), Q(19), Q(29), Q(39), Q(16), Q(18), Q(26), Q(44) for molecule 6. You should rename the listed residual electron density maxima to make hydrogen atoms out of them (do not forget to change the element identification number from 1 to 2 and to constrain the isotropic displacement to 1.2 times the value of the nitrogen atom to which the hydrogen atoms are attached) and copy them to the right location in the next .ins file (directly after the nitrogen atom to which the individual hydrogen is bound). Also include a `DFIX` restraint for each new hydrogen atom.[12] All this has been done in the file sin-05.ins.

After 20 cycles of refinement with SHELXL, the model is complete. The eight highest residual density maxima are significantly higher than the others (Q(1) = 1.24 e/Å, Q(8) = 0.99 e/Å, Q(9) = 0.48 e/Å) and are located close to the four $SiMe_3$ groups of molecules 2 and 4. It is likely that these peaks represent a second position for these $SiMe_3$ groups; however, the density maxima are too low to successfully refine a disorder (but, of course, you can always try).

The next task is to identify the non-crystallographic symmetry, if there is any. A closer look reveals that there is indeed three times twofold NCS linking molecules 1 and 5, 2 and 4, and 3 and 6: the non-crystallographic symmetry operators are one pseudo-inversion centre and two pseudo-twofold axes, one of which is only partially fulfilled, as the corresponding $SiMe_3$ groups of molecules 3 and 6 possess different torsion angles. Figure 6.2 shows the three pairs with their pseudo-symmetry

[12] The treatment of acidic hydrogen atoms is described in detail in Section 3.3.2 and in example 3.5.3.

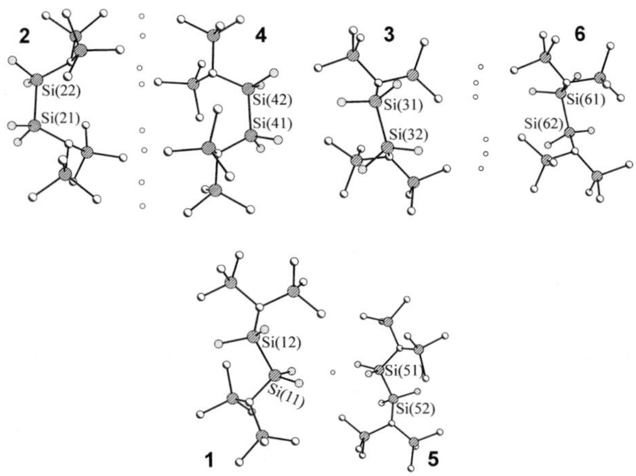

Fig. 6.2 Three times twofold non-crystallographic symmetry in the structure of $[Si(NH_2)_2CH(SiMe_3)_2]_2$. The small circles between the pairs of molecules mark the geometric centres between two NCS-related atoms. The pseudo-twofold between molecules 2 and 4 (upper left) and the pseudo-inversion centre between molecules 1 and 5 (at the bottom) are fulfilled by all atoms, while the pseudo-twofold between molecules 3 and 6 (upper right) is fulfilled only by the core atoms of the molecules but not the $SiMe_3$ ligands.

operators. Knowing which molecules are symmetry-related allows us to use similarity restraints on the 1,2- and 1,3-distances. This could be achieved by means of the `SAME` instruction as described in Chapters 2 and 5. However the resolution of the data—and hence the data-to-parameter ratio—is very good (almost 1:16), so that we might as well refrain from applying restraints.

Besides adjusting the weighting scheme, there is one more thing to do: find and describe the hydrogen bonds. As explained in Section 2.8, we add the command `HTAB` into the file sin-06.ins, which makes a table appear in the file sin-06.lst listing all independent intermolecular and intramolecular hydrogen bonds, however without standard uncertainties. Using the long form of `HTAB` in combination with the `EQIV` command (also explained in Section 2.8), we can describe every single one of the nine independent hydrogen bonds specified in this table. This has been done in file sin-07.ins and gives rise to a table of the nine hydrogen bonds with all standard uncertainties in the file sin-07.lst.

The file sin-08.res contains the final publishable model with adjusted and converged weighting scheme and `HTAB` commands for all hydrogen bonds. As you saw, the refinement of this structure was not particularly difficult. The non-crystallographic symmetry merely increased the work and computing time but did not cause a specific problem, as global pseudo-symmetry can, and if we had a resolution problem, we could even have used similarity restraints to indirectly improve the data-to-parameter ratio.

7

Twinning

Regine Herbst-Irmer

7.1 Definition of a Twin

Very often the cracking or splitting of crystals, as indicated by split reflection profiles, is loosely described as twinning. However, such crystals are just bad crystals. A working definition of a twinned crystal is the following: 'Twins are regular aggregates consisting of individual crystals of the same species joined together in some definite mutual orientation' (Giacovazzo, 2002).

To explain this definition let us assume a two-dimensional crystal (see Figure 7.1A). The content of the unit cell has mirror symmetry in one direction, while the symmetry of the cell itself (ignoring the contents), the *metric symmetry*, has an additional mirror perpendicular to the first mirror (similar to the case of a monoclinic crystal where β happens to be 90°).

If we transform this crystal by the mirror that is only fulfilled by the metric symmetry of the cell, but not by its contents, we obtain the crystal of Figure 7.1B. If both crystals grow together we have a twin (see Figure 7.2). The twin operation of this twin—the so-called *twin law*—is the mirror plane that transforms one domain into the other. As both domains in Figure 7.2 are equal in size, the fractional contributions of both domains are 0.5 and this twin is a *perfect twin*. In Figure 7.3 the fractional contributions are 0.67 : 0.33, corresponding to a partial twin.

These two features—the twin law and the fractional contribution—are necessary for the description of a twin. The twin law can be expressed as a matrix that transforms the *hkl* indices of one species into the other. If x is going down and y to right, the transformation for the cell (and the *hk* indices) in the above example would be described by the matrix: $\begin{pmatrix} 1 & 0 \\ 0 & -1 \end{pmatrix}$. Twinning may occur when a unit cell (or a supercell)—ignoring the contents, as above—has higher symmetry than implied by the space group of the crystal structure.

What happens to the diffraction pattern when twinning occurs? The crystal of Figure 7.1A would lead to a reciprocal lattice with mirror symmetry, such as can be seen in Figure 7.4A. If the crystal is transformed as in Figure 7.1B, the reciprocal lattice is also transformed (see Figure 7.4B).

If both crystals are grown together, the intensities of both reciprocal lattices are added in the twinned crystal, similar to the diffraction pattern of Figure 7.5.

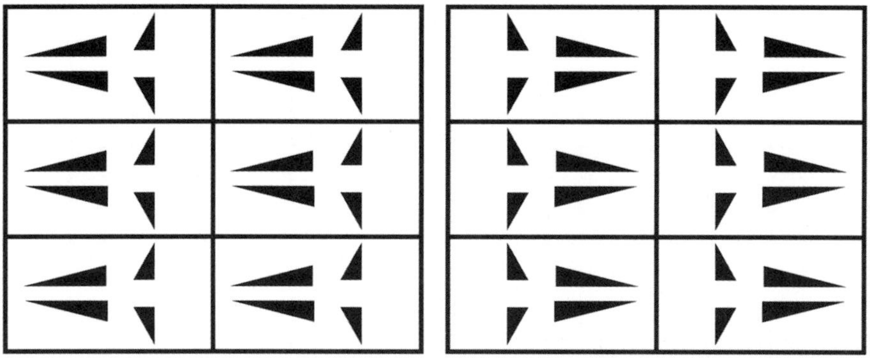

Fig. 7.1 A: Representation of a two-dimensional crystal. B: Same crystal transformed by a vertical mirror. This mirror is part of the metric symmetry of the unit cell of this crystal but not of the contents of the cell.

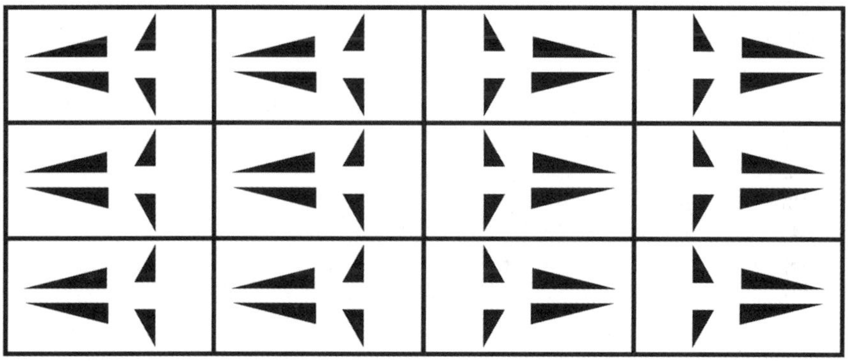

Fig. 7.2 Two-dimensional perfect twin. The two twin domains are related by the vertical mirror between the two halves of Figure 7.1.

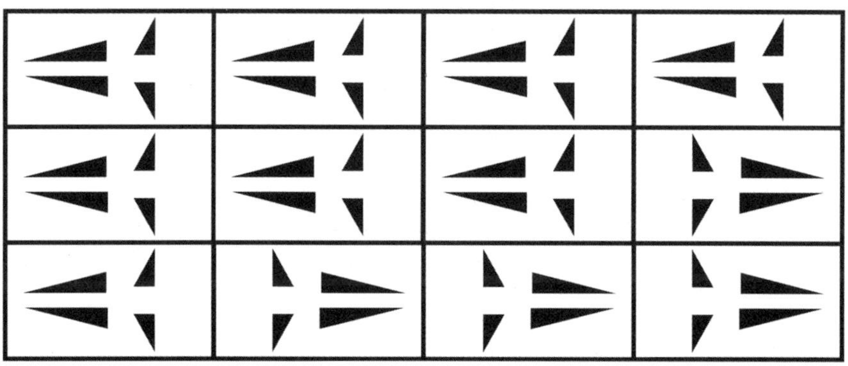

Fig. 7.3 Partial twin (ratio is 2:1) following the same twin-law as in Figure 7.2.

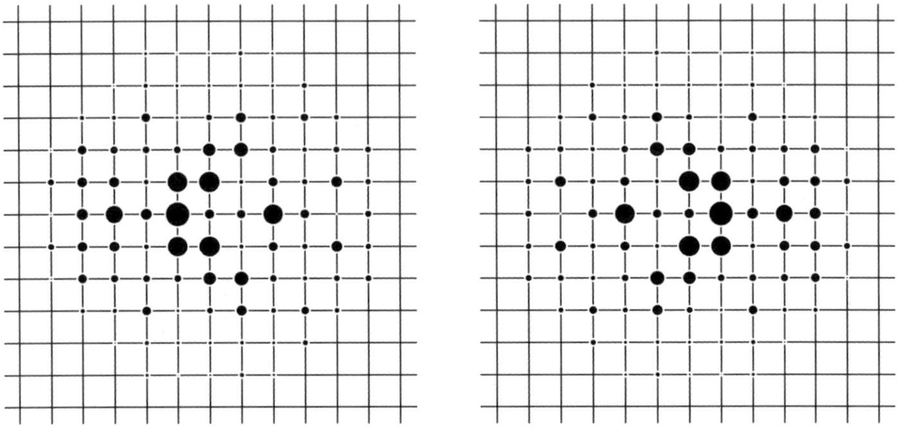

Fig. 7.4 A: Reciprocal space plot of a crystal similar to the one of Figure 7.1A. B: Reciprocal space plot of the transformed crystal (as in Figure 7.1B).

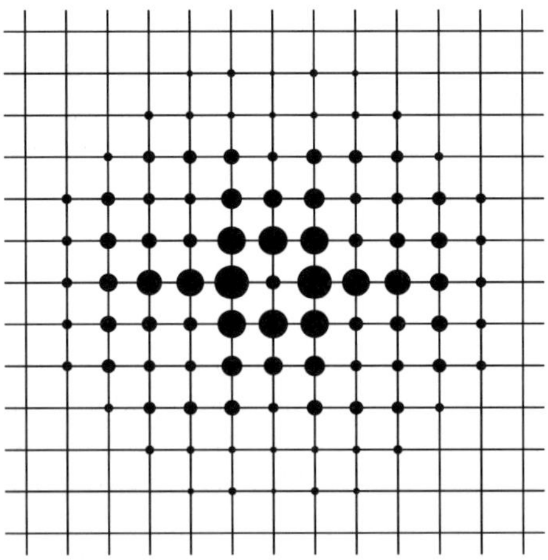

Fig. 7.5 Reciprocal space plot similar to that of the perfect twin, which can be understood as summation of the two individual diffraction patterns shown in Figure 7.4. Here we see a superposition instead of a summation. Note that additional mirror symmetry appears for this perfect twin.

7.2 Classification of twins

There are several systems of twin classification, depending for example on the morphology or on the twin element. Several types of nomenclature may also be encountered. Friedel (1928) has distinguished four kinds of twins.

7.2.1 Twinning by merohedry

In a merohedral twin, the twin law is a symmetry operator of the crystal system, but not of the point group of the crystal. This means that the reciprocal lattices of the different twin domains superimpose exactly and the twinning is not directly detectable from the reflection pattern. Two types are possible.

Racemic twins

The twin operator belongs to the Laue group but not to the point group of the crystal. These crystals are racemic twins. There are no special problems in solving and refining such structures. The only question to be resolved is the determination of the absolute structure. Even if determination of absolute configuration is not one of the aims of the structure determination, it is important to refine any non-centrosymmetric structure as the correct absolute structure or as a racemic twin, in order to avoid introducing systematic errors into the bond lengths (Cruickshank and McDonald, 1967). In some cases the absolute structure will be known with certainty (e.g. proteins), but in others it has to be deduced from the X-ray data. Generally speaking, a single phosphorus or heavier atom suffices to determine an absolute structure using Cu-$K\alpha$ radiation, and with accurate high-resolution low-temperature data including Friedel opposites, such an atom may even suffice for Mo-$K\alpha$. Of course this type of twinning cannot occur for enantiomerically pure samples of chiral molecules like protein crystals.

Other merohedral twins

The twin operator belongs to the crystal class but not to the Laue group of the crystal. This type is possible in the trigonal, tetragonal, hexagonal and cubic crystal systems, which have more than one Laue group. This type of twinning can have drastic influences on the relative intensities of the diffraction pattern and may cause severe problems (see Figure 7.6).

In Figure 7.6A a reciprocal space plot (layer $l = 0$) of a tetragonal crystal is shown. The four-fold symmetry can easily be detected, whereas there is no additional twofold axis along a^* or b^*. The Laue group is therefore $4/m$. In Figure 7.6B the reciprocal space plot is rotated about a twofold axis along a^*. Figure 7.6C shows a superposition of Figures 7.6A and 7.6B, similar to a summation resulting from twinning. Now an additional twofold axis is detectable, and the Laue group appears to be $4/mmm$. In this example both domains are equal in size (perfect twin) and the reflection intensities appear to possess a higher symmetry than the true structure. The determination of the correct space group and the structure solution can be difficult,

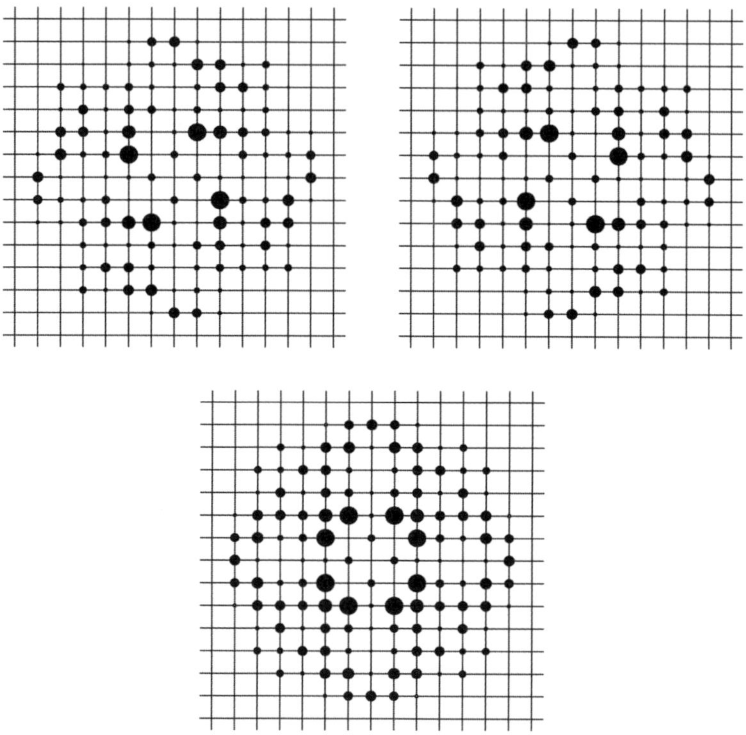

Fig. 7.6 A: Reciprocal space plot of the layer $l = 0$ (h down and k to the right) of a crystal with $4/m$ symmetry. B: Same as A but rotated by 180° about a^*. C: Superposition of the two patterns from A and B, similar to the diffraction pattern of the twinned crystal. The diffraction pattern of the perfect twin has additional symmetry, corresponding to the higher symmetry Laue group 4/*mmm*.

although SHELXD[1] is often surprisingly effective in such cases, provided that it is given the correct space group and the data were not merged using the higher apparent symmetry.

Compared to the reciprocal space plot of the untwinned crystal, the intensity distribution in the diffraction pattern of the twinned crystal has changed: in the untwinned case there are many weak and strong reflections, whereas for the twinned crystal most of the intensities lie in an intermediate range. This is because every intensity is the sum of two component intensities, whereby it will not often be the case that both intensities are large or both are small. As merohedral twinning is only possible in trigonal, hexagonal, tetragonal and cubic space groups, the number of twin laws is limited (see Table 7.1). The twin law corresponds to the twofold operation that is present in the apparent Laue group, but not in the true space group. Only for trigonal crystals is there more than one possible twin law.

[1] SHELXD is called XM in the SHELXTL world. See Chapter 1 for details.

Table 7.1 Twin laws for merohedral twins

True Laue group	Apparent Laue group	Twin law
$4/m$	$4/mmm$	0 1 0 1 0 0 0 0 −1
$\bar{3}$	$\bar{3}1m$	0 −1 0 −1 0 0 0 0 −1
$\bar{3}$	$\bar{3}m1$	0 1 0 1 0 0 0 0 −1
$\bar{3}$	$6/m$	−1 0 0 0 −1 0 0 0 1
$\bar{3}$	$6/mmm$	0 −1 0 −1 0 0 0 0 −1
		0 1 0 1 0 0 0 0 −1
		−1 0 0 0 −1 0 0 0 1
$\bar{3}m1$	$6/mmm$	−1 0 0 0 −1 0 0 0 1
$\bar{3}1m$	$6/mmm$	−1 0 0 0 −1 0 0 0 1
$6/m$	$6/mmm$	0 1 0 1 0 0 0 0 −1
$m\bar{3}$	$4\bar{3}m$	0 1 0 1 0 0 0 0 −1

Merohedral twinning may, at least in theory, occur simultaneously with racemic twinning.

7.2.2 *Twinning by pseudo-merohedry*

In a pseudo-merohedral twin, the twin operator belongs to a higher crystal system than the structure. This may happen if the metric symmetry is higher than the symmetry of the structure. Typical examples are monoclinic structures with either β very close to 90° or with a and c almost equal. Depending on how well the higher metric symmetry is fulfilled, it may happen that the reciprocal lattices overlap exactly and the twinning is not detectable from the diffraction pattern. The problems are then the same as in case of merohedral twinning: the structure appears to have a higher symmetry than it in fact possesses. Solving and refining such twins requires essentially the same procedures as for merohedral twins, but, compared to merohedral twins, the number of possible twin laws is much higher. Because of possible different settings of the true and the apparent space group, the multiplication of three matrices is sometimes necessary:

$$\begin{pmatrix} \text{apparent} \\ \downarrow \\ \text{true} \end{pmatrix} \begin{pmatrix} \text{twin operation} \\ \text{in the apparent} \\ \text{space group} \end{pmatrix} \begin{pmatrix} \text{true} \\ \downarrow \\ \text{apparent} \end{pmatrix}$$

The cell of the true space group must be transformed into the apparent cell to use the description of the twin operation in this Laue group. Then the cell must be re-transformed into the true Laue group.

In contrast to the two first types of twinning (twinning by merohedry and pseudo-merohedry), in the remaining two types not every reflection is affected by the twinning. This means that the twinning may be detectable from the diffraction pattern. In favourable cases, structure solution may be possible by identifying and using only those reflections that are contributed to by a single twin domain.

7.2.3 Twinning by reticular merohedry

A typical example is obverse/reverse twinning of a rhombohedral structure (Herbst-Irmer and Sheldrick, 2002).

For structures that crystallize in rhombohedral space groups, a twofold axis parallel to the threefold axis (matrix $-1\ 0\ 0\ \ 0\ -1\ 0\ \ 0\ 0\ 1$ in the hexagonal setting) or parallel to *a–b* (matrix $0\ -1\ 0\ \ -1\ 0\ 0\ \ 0\ 0\ -1$ in the hexagonal setting) as twin law produces a so-called obverse/reverse twin (see Figure 7.7).

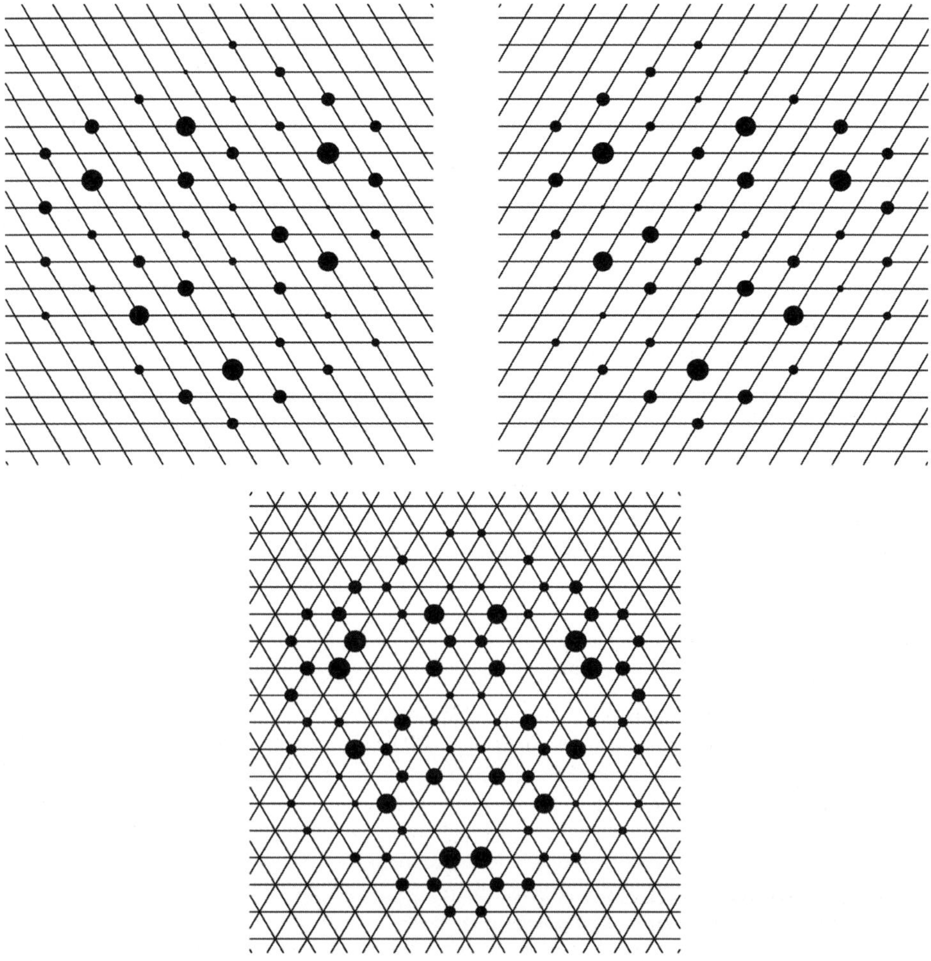

Fig. 7.7 Reciprocal space plots of the $l = 0$ layer (*h* down and *k* to the right) of a rhombohedral crystal in obverse setting (A), reverse setting (B) and the superposition of both settings (C).

In the hexagonal setting the systematic absence condition for the first domain is $-h + k + l = 3n$ (obverse setting), while for the second it is $h - k + l = 3n$ (reverse setting); it can therefore be a problem to detect this lattice centring. It can be identified by a comparison of the mean intensity or mean intensity to sigma ratio of the reflections with $-h+k+l = 3n$, $h-k+l = 3n$ and all reflections. Inspection of reciprocal space plots can also help. In layers with $l = 3n$ only every third reflection should be observed, while in all other layers one third of the reflections are absent (see Figure 7.7C). Version 6.12 (and higher) of the program XPREP (Sheldrick, 2001) gives further help: it checks and compares the mean intensity for reflections

(1) that should be observed only in case of the obverse setting
(2) that should be observed only in case of the reverse setting
(3) that should be absent in both cases.

Then it estimates the fractional contributions of the second domain.

With obverse/reverse twinning there are four types of reflections: reflections with $-h + k + l = 3n$ and $h - k + l \neq 3n$ are only observed for the main domain, reflections with $-h + k + l \neq 3n$ and $h - k + l = 3n$ have non-zero intensity only for the second domain, reflections with $-h + k + l \neq 3n$ and $h - k + l \neq 3n$ are absent for both domains, and reflections with $-h + k + l = 3n$ and $h - k + l = 3n$ have contributions from both domains. Because only one third of the reflections (those with $l = 3n$) are affected by the twinning, structure solution is normally not a severe problem, because two thirds of the reflections have contributions from only one domain and are often sufficient for structure solution.[2]

For the refinement of an obverse/reverse twin SHELXL needs a special reflection file in HKLF 5 format and the refinement is not possible with a single TWIN command (see the two examples in 7.8.3 and 7.8.4). This restriction is unnecessary and will be removed if and when there is a new release of the program. After producing the HKLF5 format file further merging of equivalent reflections is not possible. Therefore the data should be merged before producing this file. Otherwise all data would be treated as independent, which leads to mathematically incorrect standard uncertainties.

Generation of the HKLF 5 file

Reflections that are absent for both domains are omitted, and in practice it may well be expedient to omit also those reflections that have only a contribution from the second domain.[3] Reflections that have a contribution only from the main domain are unchanged and are assigned the batch number 1. Reflections with contributions from both domains are split into their two components $-h - kl$ and hkl (if the twin axis is parallel to c) or $-k - h - l$ and hkl (if the twin axis is perpendicular to c)[4]

[2] XPREP is able to produce a crude untwinned data set, if more data are required for structure solution. Untwinned data should however never be used for the final refinement because of correlations between twin-related reflections.

[3] Usually the second domain is weaker and is often not as well centred in the beam. These additional data are thus of poorer quality and often do not improve the model. Secondly they would be treated as independent data, but are of course not independent of reflections of the first domain with the same indices, which would tend to falsify the standard uncertainties.

[4] In the higher symmetry trigonal Laue group these two twin laws are equivalent.

and are assigned the batch numbers −2 for contribution of the second component or 1 for the contribution of the first component. The batch numbers −2 and 1 tell the program that these two reflections of domains 2 and 1 contribute to one observed intensity. Only the last reflection in a group of overlapping reflections is given a positive batch number.

For structures crystallizing in the lower symmetry rhombohedral Laue group, in addition to obverse/reverse twinning, the twofold axis parallel to a may act as a further twin law (matrix 0 1 0 1 0 0 0 0 −1). In this case the twinned reflection data file will contain up to four contributions to each observed intensity. Reflections that are only present for the obverse setting contain the two components $kh - l$ and hkl with batch numbers −3 and 1, while reflections with $l = 3n$ contain the four components: $-k - h - l$, $kh - l$, $-h - kl$, and hkl with the batch numbers assigned as −4, −3, −2, and 1 (see the second example in 7.8.3).

7.2.4 Non-merohedral twins

For non-merohedral twins, the twin law does not belong to the crystal class of the structure nor to the metric symmetry of the cell. Therefore the different reciprocal lattices do not overlap exactly. There are some reflections, which overlap or cannot be distinguished from each other, but the majority of the reflections are not affected by the twinning. As shown in Figure 7.8, the diffraction pattern should normally reveal this type of twinning.

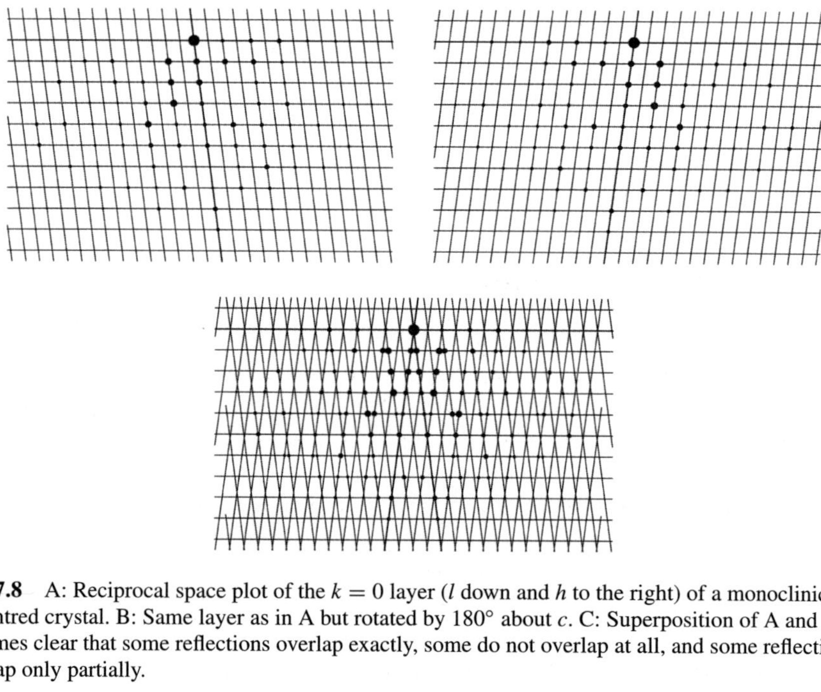

Fig. 7.8 A: Reciprocal space plot of the $k = 0$ layer (l down and h to the right) of a monoclinic C-centred crystal. B: Same layer as in A but rotated by 180° about c. C: Superposition of A and B. It becomes clear that some reflections overlap exactly, some do not overlap at all, and some reflections overlap only partially.

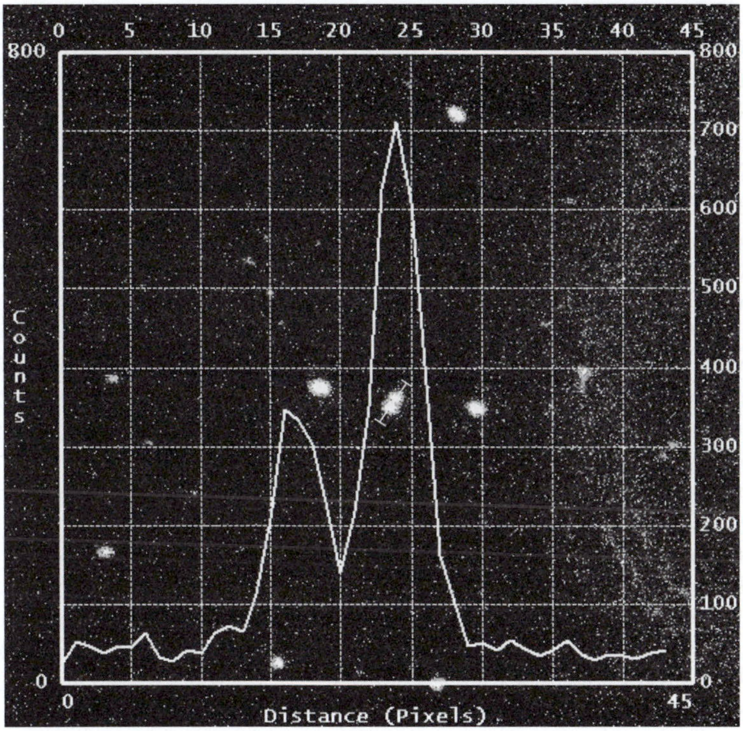

Fig. 7.9 Typical diffraction pattern of a non-merohedral twin. Next to normal looking reflections there is one reflection, which is split. The intensity profile has been drawn along a line across this reflection.

Normally this kind of twinning is detected at the diffractometer, because the automatic cell determination programs fail or have problems. Split reflection profiles or un-indexed reflections are typical warning signs (Figure 7.9).

Unit cell determination and determination of the twin law

In case of a two-domain non-merohedral twin, two orientation matrices must be determined. The programs DIRAX (Duisenberg, 1992) and GEMINI (Sparks, 1997; Bruker-AXS, 1999) take into account that only a certain fraction of the reflections will fit an individual solution. As a result, a list of possible solutions is presented. If a solution is accepted, all the reflections that do not fit this first solution are placed in a new reflection list to re-run the cell determination process. After the determination of both orientation matrices the twin law can be calculated in a separate step:

$$T = A_2^{-1} \cdot A_1 \tag{7.1}$$

$$T \cdot h_1 = h_2 \tag{7.2}$$

with A_i orientation matrix i, h_i indices i and T the twin law.

If the second domain is much weaker, the determination of the orientation matrix of this domain could be problematic, because there are only a few strong reflections that fit only this weak domain. For this reason, the program CELL_NOW (Sheldrick, 2003), developed especially for problematic cases, has a different approach to derive several orientation matrices. It tries to find sets of reciprocal lattice planes that pass close to as many reflections as possible. A figure of merit describes the goodness of the fit. CELL_NOW tries to find the best cell within a given cell length range taking several figures of merit into account (e.g. small volume, high symmetry, high fraction of reflections that fit). The reflections that fit this cell within a specified fraction of all three interplanar spacings may be flagged as indexed. Instead of trying to index the remaining reflections with a completely independent orientation matrix, CELL_NOW uses the cell information; it rotates the first cell to locate further twin domains iteratively, using only the reflections that have not yet been indexed. The rotation matrix is then the twin law. Thus the orientation matrices and the twin law are determined in one step.

Sometimes the twinning is not detected at the diffractometer. The second domain is so weak that the cell determination proceeds without any severe problem. But then the refinement will be unsatisfactory. For such cases the program ROTAX (Cooper *et al.*, 2002) was developed. It checks a list of those reflections with the greatest differences between F_o and F_c and for which F_o is bigger than F_c. It assumes that the additional intensity is produced by a second twin domain. ROTAX produces a list of possible rotations and tests for each of them whether this rotation would lead to an overlap for the reflections of its list. For this purpose it calculates the transformation matrix for the Bragg indices for every rotation and checks then the deviation from integer values of the resulting indices of all reflections of the reflection list. If the mean deviation is low the rotation matrix is a possible twin law. A similar procedure is programmed in TwinRotMat (Spek, 2006).

Data processing

In the general case of a non-merohedral twin, there are three different types of reflections: reflections that do not overlap with any reflection of the second domain, reflections that overlap exactly with a reflection of the second domain, and finally reflections that overlap partially with a reflection of the second domain. The last type is the most problematic, because usually the degree of overlap is not known and differs from reflection to reflection. When only one orientation matrix is used to integrate the data, part of the intensity of the reflection of the second domain is added to the intensity of the reflection of the main domain.

Using the second orientation matrix, it is possible to integrate the whole intensity of the overlapping reflection. Such an integration is available with SAINT (Bruker, 2001) or EvalCCD (Duisenberg *et al.* 2003).

The procedure for SAINT is as follows: the program checks whether there is an overlap and tries to integrate the intensity of the individual reflections. The raw reflection file consists of non-overlapped reflections and overlapped reflections split into their different components. To distinguish it from the standard raw-files the file extension is changed to .mul.

Absorption correction, scaling, merging and generation of twin reflection file
As the format of this new type of raw file is different from the standard SAINT output files, .mul files need a special version of the absorption correction program SADABS (Sheldrick, 1997a), called TWINABS (Sheldrick, 2002). Additionally to the scaling and absorption procedure, TWINABS produces a special twin reflection file (HKLF5 format file) that describes in detail which reflections are affected by overlap.

There are different options in the refinement of parameters to model systematic errors, for example using all reflections for one model for all domains, using only reflections of one domain or using separate models for different domains. Normally one model for all domains is appropriate, but if both domains are macroscopically distinguishable and are similar in size, two different error models generated only by non-overlapping reflections could be superior—provided there are enough data.

For the output there are also various possibilities: detwinned (HKLF4 format file for structure solution) or twinned (HKLF5 format file), merged or unmerged, and all reflections or only reflections with the contribution of one domain (the latter option only for HKLF5 format file). After producing the special format for twinned data (SHELXL HKLF5 format), further merging of equivalent reflections in SHELXL is not possible. Therefore the reflections should be merged before producing this file. Otherwise all data would be treated as independent, which leads to an artificially large number of data.

In general, only reflections with a contribution from the main domain should be used, because in most cases this domain is much better determined. Even if data of both domains are similar in quality, the addition of reflections having only a contribution from the second domain does not usually improve the structure determination. Again, this procedure would artificially increase the number of apparently independent data.

Remaining problems
The critical point of this integration method is to determine whether there is overlap between two reflections or not. As will be shown with the example in 7.8.4, there is a relatively large number of reflections for which SAINT assumes no overlap, while there are symmetry-related or even identical reflections collected on different frames which SAINT treats as overlapping. This prevents TWINABS from merging these reflections, which results in an artificially high number of independent reflections. This is generally thought to lead to mathematically incorrect standard uncertainties though the differences are usually small. In principle it could be possible to have an overlap for one reflection but no overlap of a symmetry-related reflection, associated with different measuring positions, but in our experience this is rare. However, in our

example structures, the number of cases of inconsistent overlap is about 20% or even higher, which seems to be too high to ignore. Very often these reflections appear to fit worst to the model. So a simple approach would be to omit these unsplit reflections where a symmetry related one is overlapped, while the overlapped reflection is used in the refinement. The examples show that this often leads to small improvements, but the differences are so small that they need not necessarily be taken into account for routine structure determination, especially if the redundancy is high enough.

7.3 Tests for twinning

As mentioned above (pseudo-)merohedral twinning is not detectable from the reflection pattern, but the intensity distribution of twins is different from that of untwinned crystals. This phenomenon is used in several tests for twinning.

The program XPREP uses the mean value of $|E^2 - 1|$. E-values are normalized F-values, and the expected mean value for E^2 is thus 1. The mean value of $|E^2 - 1|$ is the variance. The theoretically expected value for centrosymmetric structures is 0.968, that for non-centrosymmetric structures 0.736. A higher value means that the difference in intensities is higher, so there are many weak and also many strong reflections. For (pseudo-)merohedrally twinned structures $\langle|E^2 - 1|\rangle$ may be much lower than the expected values. Because every intensity is the sum of two intensities, as explained above, it will very seldom happen that both intensities are high or both are small. Additionally XPREP compares the R_{int}-values of the possible Laue groups (for a definition of R_{int} see Chapter 1). For a partially twinned structure the R_{int} for the apparent Laue group should be only slightly higher than for the correct one. The difference in R_{int} values is dependent on the fractional contribution.

Other tests based on intensity statistics have been proposed (e.g. see Rees, 1980; Yeates, 1997 (www.doe-mbi.ucla.edu/Services/Twinning); Kahlenberg, 1999; Kahlenberg and Messner, 2001). Nevertheless, the use of the mean $|E^2 - 1|$ is particularly simple, because it is a single number and is often calculated by data reduction and direct methods programs.

There are some phenomena other than twinning that affect the differences in intensities, such as anisotropic data or translational pseudo-symmetry. Anisotropic data is a problem more frequently observed for protein structures, while translational pseudo-symmetry may also occur in small molecules. Consider a structure with a small number of heavy metal atoms and a greater number of carbon atoms. It may happen that the metal atoms fulfil a lattice centring, for example a C-centring, whereas the carbon atoms correspond to a primitive lattice. Because of the high contribution of the metal atoms to the scattering power, all reflections with $h + k \neq 2n$ will be very weak, because only the carbon atoms contribute to them. So the spread in intensities will be higher than for a random distribution of atoms. In such a case $\langle|E^2 - 1|\rangle$ will be higher than expected. If this crystal is then (pseudo-) merohedrally twinned it might happen that $\langle|E^2 - 1|\rangle$ has a normal value, because both effects cancel. If then additionally the fractional contributions of both twin domains are

Table 7.2 Expected values for the function L

	$\langle \lvert L \rvert \rangle$	$\langle L^2 \rangle$
Acentric reflections of an untwinned crystal	1/2	1/3
Centric reflections of an untwinned crystal	$2/\pi$	1/2
Acentric reflections of a perfectly twinned crystal	3/8	1/5

similar in size, XPREP cannot decide between a perfect twin in the lower symmetry Laue group and an untwinned structure in the higher symmetry Laue group. This is also true for most of the other intensity tests. Padilla and Yeates (2003) proposed a test that seems to overcome this problem, but it is mainly intended for protein structures. In the function L, the intensities I of neighbouring reflections h_1 and h_2 are used. Table 7.2 summarizes expected values for this function:

$$L \equiv \frac{I(h_1) - I(h_2)}{I(h_1) + I(h_2)} \tag{7.3}$$

7.4 Structure solution

As mentioned above, structure solution may be difficult for twins where every reflection is affected by the twinning, and especially for those with similar domain sizes. For small molecules, normal direct methods are often able to solve twinned structures even for perfect twins, provided that the correct space group is used. The program SHELXD is even able to utilize the twin law and the fractional contribution (Usón and Sheldrick, 1999).

The Patterson function of a twinned structure is the sum of the Patterson functions of both domains. Therefore procedures using Patterson methods are in principle possible. There are several examples in the literature of structures solved by molecular replacement using twinned data (e.g. see Breyer *et al.*, 1999).

For partially twinned structures, mathematical detwinning is possible if the fractional contribution is not too close to 0.5. The intensities J_1 and J_2 measured from a twinned crystal are the sum of the two intensities I_1 and I_2 of both domains weighted by their fractional contribution α:

$$J_1 = (1-\alpha)I_1 + \alpha I_2 \tag{7.4}$$

$$J_2 = \alpha I_1 + (1-\alpha)I_2 \tag{7.5}$$

Thus, the contributions of the two domains can be calculated, assuming $\alpha \neq 0.5$

$$I_1 = \frac{(1-\alpha)J_1 - \alpha J_2}{1-2\alpha} \tag{7.6}$$

$$I_2 = \frac{(1-\alpha)J_2 - \alpha J_1}{1-2\alpha} \tag{7.7}$$

This detwinned data can be used for structure solution, whereas refinement should be performed against the original data, because for α approaching 0.5 detwinned intensities become very inaccurate.

There are also examples of structures solved by MAD/SAD using twinned or detwinned data (e.g. see Rudolph *et al.*, 2003). However, care should be exercized in detwinning anomalous diffraction data in order to avoid mixing the positive and negative Friedel mates (Dauter, 2003).

7.5 Twin refinement

In SHELXL the twin refinement method of Pratt *et al.* (1971) and Jameson (1982) has been implemented. F_c^2 values are calculated by:

$$\left(F_c^2\right)^* = \text{osf}^2 \sum_{m=1}^{n} k_m F_{c_m}^2 \tag{7.8}$$

where osf is the overall scale factor, k_m is the fractional contribution of twin domain m and F_{c_m} is the calculated structure factor of twin domain m. The sum of the fractional contributions k_m must be unity, so $(n - 1)$ of them can be refined and k_1 is calculated by:

$$k_1 = 1 - \sum_{m=2}^{n} k_m \tag{7.9}$$

For completely overlapping lattices, a normal intensity data file (standard HKLF4 format) can be used together with the following two instruction lines:

```
TWIN r11 r12 r13 r21 r22 r23 r31 r32 r33 n
BASF k2 k3 ... kn
```

The matrix *rij* is the twin law and n the number of twin domains. The batch scale factor `BASF` is followed by $n-1$ starting values for the fractional contributions. The default value for n is 2, which corresponds to a twin with two domains.

If only part of the reflections have a contribution from the second domain (twinning by reticular merohedry and non-merohedral twins), a special reflection file is necessary, which is read in by the command

```
HKLF 5
```

The `HKLF 5` is given at the end of the .ins file, replacing the line that reads `HKLF 4`. `BASF` is used as in the case before. As merging is no longer allowed the default value for `MERG` assumed by SHELXL is now 0.

Generally, twinned crystals tend to have a poor effective data to parameter ratio, so they often require restraints in order to obtain a satisfactory refinement (Watkin, 1994). The following restraints can be useful: distance restraints for chemically equivalent 1,2- and 1,3-distances, planarity restraints for groups such as phenyl rings, rigid bond ADP restraints (Hirshfeld, 1976; Rollett, 1970; Trueblood and Dunitz, 1983) and 'similar ADP restraints' (Sheldrick, 1997b).[5] Even when restraints are employed, the distribution of the displacement parameters (ORTEP plot) and residual features in a difference electron density map can be less satisfactory than for a normal structure determination.

7.6 Determination of the absolute structure

The definition of the Flack parameter (Flack, 1983; Bernadinelli and Flack, 1985) is a special case of Equation 7.8:

$$\left(F_c^2\right)^* = (1 - x)F_c^2(hkl) + xF_c^2(-h - k - l) \tag{7.10}$$

Here x is the fractional contribution of the inverted component of an assumed racemic twin. It should be zero if the absolute structure is correct, unity if it has to be inverted, and somewhere in between if racemic twinning is genuinely present (note that the value can only be judged together with its standard uncertainty). Thus the above formulae apply with $n = 2$ and the twin law $R = (-1\,0\,0, 0\,-1\,0, 0\,0\,-1)$. This matrix is the default matrix for `TWIN`, therefore the two following commands refine racemic twins.

```
TWIN
BASF k2
```

7.7 Warning signs of twinning

Experience shows that there are a number of characteristic warning signs of twinning, as given in the following list (Herbst-Irmer and Sheldrick, 1998). Of course not all of them can be present in any particular example, but if one or several apply, the possibility of twinning should be given serious consideration.

1. The metric symmetry is higher than the Laue symmetry.
2. The R_{int}-value for the higher symmetry Laue group is only slightly higher than for the lower symmetry Laue group.
3. If different crystals of the same compound show significantly different R_{int} values for the higher symmetry Laue group, this clearly shows that the lower symmetry Laue group is correct and indicates different extents of twinning.
4. The mean value for $|E^2 - 1|$ is much lower than the expected value of 0.736 for the non-centrosymmetric case. If we have two twin domains and every reflection

[5] For a detailed description of the restraints see Chapter 1.

has contributions from both, it is unlikely that both contributions will have very high or that both will have very low intensities, so the combined intensities are distributed to give fewer extreme values.

5. The space group appears to be trigonal or hexagonal.
6. The apparent systematic absences are not consistent with any known space group.
7. Although the data appear to be in order, the structure cannot be solved. This may of course also happen if the cell is wrong, for example with an halved axis.
8. The Patterson function is physically impossible.

The following features are typical of non-merohedral twins, where the reciprocal lattices do not overlap exactly and only some of the reflections are affected by the twinning:

9. There appear to be one or more unusually long axes.
10. There are problems with the unit cell refinement.
11. Some reflections are sharp, others split.
12. $K = \mathrm{mean}(F_o^2)/\mathrm{mean}(F_c^2)$ is systematically high for reflections with low intensity. This may also indicate a wrong choice of space group in the absence of twinning.
13. For all of the 'most disagreeable reflections' in the .lst file, F_o is much greater than F_c.
14. Strange residual electron density, which cannot be resolved as solvent or disorder.
15. High R-values although the data seem to be of good quality. Of course there are many more possible explanations for this phenomenon.

7.8 Examples

In the following sections we present examples of how to refine twinned structures with SHELXL. All files you may need in order to perform the refinements yourself are given on the CD-ROM that accompanies this book. The first example is a case of merohedral twinning that will acquaint you with the basics of practical twin refinement. The second example describes a typical pseudo-merohedral twin such as every crystallographer will encounter sooner or later. Two different examples for twinning by reticular merohedry are given next and the chapter ends with two cases of non-merohedral twinning.

7.8.1 Twinning by merohedry

The first structure (Herbst-Irmer and Sheldrick, 1998) is an example of twinning by merohedry, which could not be solved by routine methods. The composition of the compound was not known with certainty, but an osmium compound with some triphenylphosphine and chloro ligands was expected. The first problem is to determine the space group. XPREP gives the following output (mero.prp):

```
Original cell in Angstroms and degrees:

  12.623  12.623  26.325  90.00  90.00  120.00

 6579 Reflections read from file mero.hkl; mean (I/sigma) = 12.17

SPACE GROUP DETERMINATION

Lattice exceptions: P    A     B     C     I     F    Obv   Rev   All

N (total) =         0  3280  3280  3282  3281  4921  4379  4382  6579
N (int>3sigma) =    0  2552  2535  2579  2568  3833  3428  3416  5121
Mean intensity =  0.0  78.2  76.9  76.8  76.4  77.3  75.5  73.8  75.9
Mean int/sigma =  0.0  12.3  12.3  12.3  12.3  12.3  12.3  12.0  12.2

Crystal system H and Lattice type P selected

Mean |E*E-1| = 0.510 [expected .968 centrosym and .736 non-centrosym]

Chiral flag NOT set

Systematic absence exceptions:


        61/65 62=31  63    -c-   --c

N          22    17    14   464   266
N I>3s      5     0     5   333   215
<I>     258.0   4.7 403.4 106.8 106.8
<I/s>    16.7   0.8  25.9  16.1  16.4


Identical indices and Friedel opposites combined before calculating R(sym)

Option  Space Group  No. Type Axes CSD R(sym) N(eq) Syst. Abs.      CFOM


[A] P3(1)        #144  chiral  1   68   0.067   2278  0.8 / 12.2    7.38
[B] P3(2)        #145  chiral  1   68   0.067   2278  0.8 / 12.2    7.38
[C] P3(1)21      #152  chiral  1   82   0.120   4108  0.8 / 12.2   30.65
[D] P3(2)21      #154  chiral  1   82   0.120   4108  0.8 / 12.2   30.65
[E] P3(1)12      #151  chiral  1    2   0.318   4238  0.8 / 12.2  190.82
[F] P3(2)12      #153  chiral  1    2   0.318   4238  0.8 / 12.2  190.82
[G] P6(2)        #171  chiral  1    6   0.286   4364  0.8 / 12.2  155.21
[H] P6(4)        #172  chiral  1    6   0.286   4364  0.8 / 12.2  155.21
[I] P6(2)22      #180  chiral  1    9   0.323   5216  0.8 / 12.2  204.06
[J] P6(4)22      #181  chiral  1    9   0.323   5216  0.8 / 12.2  204.06
```

The crystal appears to be trigonal with $a = b = 12.623(2)$ and $c = 26.325(5)$ Å. There are systematic absences for a 3_1 or 3_2 axis. The R_{int} value for the Laue group $\bar{3}$ is quite acceptable (0.067), but the value for the Laue group $\bar{3}m$ is only slightly higher (0.120).[6] We decide on space group $P3_2$ and set up the file mero01.ins for solution with Patterson methods (`PATT` instruction instead of `TREF`). It is possible to obtain the coordinates of the osmium and of four phosphorus or chlorine atoms from the Patterson function in the space group $P3_2$ (see files mero01.res and mero01.lst). In the .lst file you can find the following crossword table:

```
Solution  1   CFOM = 38.99   PATFOM = 82.0   Corr. Coeff. = 69.0  SYMFOM = 99.9

Shift to be added to superposition coordinates:  0.0579  0.1868  0.0000


Name  At.No.  x     y       z     s.o.f. Minimum distances / PATSMF (self first)

OS1 81.2 0.8829 0.6197 0.5000 1.0000 11.46
                                     179.3

CL2 21.3 0.7400 0.4227 0.5001 1.0000 8.96  2.23
                                      2.3  30.6

CL3 20.2 1.2309 0.7556 0.5003 1.0000 9.49  3.84  5.48
                                     16.0  12.6   0.0

CL4 17.1 1.0726 0.9680 0.4981 1.0000 9.01  3.81  8.64  4.07
                                      6.2  20.9   0.0   9.7

P5  14.8 0.7861 0.6265 0.5885 1.0000 10.40 2.65 3.30  5.52 4.66
                                       5.1 20.6  0.0  0.0  0.0

P6  14.7 1.0460 0.6141 0.5349 1.0000 10.51 2.29 3.50 2.30 4.42 3.65
                                      68.7 20.8  2.8  0.0  0.0  0.4

P7  13.9 1.2088 0.7667 0.5446 1.0000 9.75  3.76 5.44 1.23 3.91 4.85 2.01
                                      2.6  13.8  0.0  0.9  0.0  0.9  0.0

P8  13.4 0.9485 0.8371 0.4976 1.0000 9.32 2.44 4.53 4.18 1.61 3.40 3.72 4.00
                                      0.0 21.2  2.8  0.0  0.0  0.0  2.0  0.0
```

The boldface atoms display a reasonable geometry, and most of the Patterson minimum function values differ from 0 (for interpretation of such 'crossword tables' see: Sheldrick, 1992). We are keeping those five atoms (Os(1), Cl(2), P(5), P(6), and P(8)) and generate the file mero02.ins (give `PLAN 100` to generate sufficiently many residual density maxima). Refining these atoms produces a difference electron

[6] For a definition of R_{int} see Chapter 1.

density map that is not very satisfactory. In spite of the relatively low R-values ($wR2$(all data) $= 0.57$, $R1(F > 4\sigma(F)) = 0.24$)[7] only a small part of the structure can be identified (see file mero02.res). However there were some typical warning signs of twinning. The mean value of $|E^2 - 1|$ is very low (0.510) and the R_{int} value for the higher symmetric Laue group is significantly, but only slightly, higher than for the lower symmetric one. This could mean that the twofold axis is not a true crystallographic axis but the twin law, so the matrix is 0 1 0 1 0 0 0 0 −1.

The twinning test with XPREP confirms this hypothesis (see file mero.prp):

```
Comparing true/apparent Laue groups.  0.05 < BASF < 0.45 indicates
partial merohedral twinning.  BASF ca. 0.5 and a low <|E^2-1|>
(0.968[C] or 0.736[NC} are normal) suggests perfect merohedral
twinning.  For a twin, R(int) should be low for the true Laue group
and low/medium for the apparent Laue group.

[1] -3 / -31m: R(int) 0.067(2278)/0.335(1960), <|E^2-1|> 0.499/0.366
TWIN  0 -1 0  -1 0 0  0 0 -1       BASF  0.253 [C]  or  0.186 [NC]

[2] -3 / -3m1: R(int) 0.067(2278)/0.124(1830), <|E^2-1|> 0.499/0.475
TWIN  0 1 0  1 0 0  0 0 -1         BASF  0.321 [C]  or  0.272 [NC]

[3] -3 / 6/m: R(int) 0.067(2278)/0.321(2086), <|E^2-1|> 0.499/0.374
TWIN  -1 0 0  0 -1 0  0 0 1        BASF  0.196 [C]  or  0.113 [NC]

[4] -31m / 6/mmm: R(int) 0.335(1960)/0.110(978), <|E^2-1|> 0.366/0.354
TWIN  -1 0 0  0 -1 0  0 0 1        BASF  0.364 [C]  or  0.326 [NC]

[5] -3m1 / 6/mmm: R(int) 0.124(1830)/0.357(1108), <|E^2-1|> 0.475/0.355
TWIN  -1 0 0  0 -1 0  0 0 1        BASF  0.254 [C]  or  0.186 [NC]

[6] 6/m / 6/mmm: R(int) 0.321(2086)/0.125(852), <|E^2-1|> 0.374/0.361
TWIN  0 1 0  1 0 0  0 0 -1         BASF  0.380 [C]  or  0.347 [NC]
```

To take this twinning into account enter the two lines TWIN 0 1 0 1 0 0 0 0 −1 and BASF 0.4 in the file mero02.ins and save it under the name mero03.ins. This simple change substantially improves the refinement (see files mero03.res and mero03.lst). The R-values drop to 0.13 ($R1$) and 0.35 ($wR2$) and several phenyl rings can now be located.

After only a few more cycles of refinement the whole structure can be found. The final model corresponds to the file mero04.res. It should be mentioned that electron density maps of twinned structures at the beginning of refinement are not as clear-cut as those of normal single crystals. Often more intermediate steps are necessary to complete the structure.

[7] For a definition of the R-values see Chapter 1.

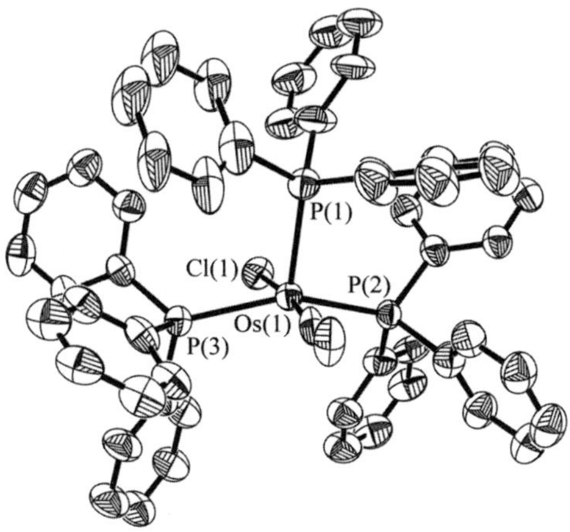

Fig. 7.10 Final structure[8] corresponding to the file mero06.res. The disordered ethanol molecule and hydrogen atoms have been omitted for clarity.

It was necessary to introduce some restraints: there are nine chemically equivalent phenyl rings, so all chemically equivalent 1,2- and 1,3-distances in the nine rings are restrained to be the same. For every phenyl ring a planarity restraint is employed. For the anisotropic displacement parameters of the carbon atoms it was necessary to use the rigid bond and similarity restraints. There is also one disordered ethanol molecule in the cell; distance and ADP restraints are employed to refine it.

The refinement of the fractional contribution k_2 converges to 0.393(2), $R1$ to 0.0547 and $wR2$ to 0.1348 (see mero04.lst). Figure 7.10 shows the final structure.

In acentric space groups and in the presence of heavy atoms such as osmium it should be possible to determine the absolute structure and the absolute structure parameter needs to be checked (.lst file). The Flack parameter x (Flack, 1983) is refined to 0.54(2).[9] This could mean that the absolute structure is wrong and space group $P3_1$ is the correct one instead of $P3_2$ and/or that there is some additional racemic twinning. This can be tested by changing the `TWIN` and `BASF` command lines:

```
TWIN 0 1 0 1 0 0 0 0 -1 -4
BASF 0.2 0.2 0.2
```

The 4 tells the program that there are now four twin domains and the minus sign means that racemic twinning should be taken into account. Because of the four twin domains three values are needed on the `BASF` card. These changes have been made in the file mero05.ins, and after ten cycles of refinement the fractional contributions

[8] There is still a hydrogen missing bond to Osmium, which cannot be found unambigously.

[9] This is the reason for the warning generated by SHELXL on the screen: 'Possible racemic twin . . .'

refine to the following values:

```
N       value          esd      shift/esd  parameter

1      0.26977      0.00054      -0.001      OSF
2      0.53534      0.02520      -0.003    FVAR   2
3      0.31975      0.01516       0.000    BASF   1
4      0.56352      0.01945       0.000    BASF   2
5      0.07401      0.01514       0.000    BASF   3
```

`BASF 1` is the fractional contribution k_2 of the second domain generated by the operation of the twofold axis with the matrix 0 1 0 1 0 0 0 0 -1, `BASF 2` is the fractional contribution k_3 of the third domain with the operation of the inversion centre and the matrix -1 0 0 0 -1 0 0 0 -1 and `BASF 3` is the fractional contribution k_4 of the fourth domain after applying both operations for the twofold axis and the inversion centre, the mirror with the matrix 0 -1 0 -1 0 0 0 0 1. The refined value for k_1 ($k_1 = 1 - (k_2 + k_3 + k_4)$) is very close to 0. This means that we do not have the original domain but the domain with the operation of an inversion centre. So we have the wrong absolute structure and must therefore invert the structure. In this case this also means that we must change the space group from $P3_2$ to $P3_1$. Inverting the coordinates can be done with the command

```
MOVE 1 1 1 -1
```

The first three numbers are added to the fractional coordinates x, y and z, respectively. The fourth number multiplies all three coordinates x, y, and z. So the above command changes x, y, z into $-(1 + x)$, $-(1 + y)$, $-(1 + z)$.

Also the refined value for k_4 is very close to zero. So we do not have four domains but two. The components 2 and 3 with highest twin fractions have matrices related by a mirror plane, not a twofold axis, so the twin law has to be changed from a twofold axis into a mirror plane with the matrix 0 -1 0 -1 0 0 0 0 1. All necessary changes have been made in the file mero06.ins. Now the refinement (see mero06.res and mero06.lst) is satisfactory: $R1 = 0.049$, $wR2 = 0.122$, $k_2 = 0.394(2)$, Flack $x = 0.03(2)$; especially taking into account that the data were collected years ago on a four-cycle-diffractometer with a proportional counter and at room temperature.

7.8.2 An example of pseudo-merohedral twinning

The following data set of aniline (Gornitzka, 1997 personal communication) was collected at -100°C and there were no problems in the integration process. The space group was determined to be $P2_1/c$ (see file pmero.prp). The file pmero01.ins on the accompanying CD-ROM corresponds to the SHELXS input for this structure. The structure can be easily solved by direct methods (see pmero01.res), and all atoms can be found. However the refinement only converges to $R1$ 0.071 (corresponding to pmero02.res), although the data appear to be of better quality. Additionally the refinement statistics show some strange features: in the file pmero02.lst you can find the following table:

```
Analysis of variance for reflections employed in refinement
K = Mean[Fo^2] / Mean[Fc^2]   for group

Fc/Fc(max)      0.000 0.009 0.017 0.026  0.036  0.047  0.061 0.077 0.104  0.152  1.000
Number in group    197.   164.   178.    188.    173.    189.   163.   182.    178.    178.
           GooF  1.664 1.428 1.579  1.612  1.174  0.867 0.926 0.898  0.916  1.530
            K    6.814 1.807 1.486  1.246  1.096  1.009 1.017 1.008  1.004  1.021
```

In this analysis of variance for reflections employed in refinement, the reflections are grouped depending on their intensity. For each group GooF and *K*, which is defined as mean[F_o^2] / mean[F_c^2], are determined. Here *K* differs significantly from unity for the reflections with the lowest intensities. This could mean that there is some extra intensity caused by a second twin domain. When we look at the diffraction data more carefully (go back to the original XPREP output), the LePage Algorithm (LePage, 1982) implemented in XPREP results in the following output (see file pmero.prp):

```
Search for higher metric symmetry
Identical indices and Friedel opposites combined before calculating R(sym)
--------------------------------------------------------------------------------
Option A: FOM = 0.041 deg.    ORTHORHOMBIC C-lattice    R(sym) = 0.327 [  2483]
Cell:     8.319  42.477   5.833   90.00   90.00   90.04    Volume:      2061.20
Matrix: 0.0000  0.0000  1.0000  2.0000  0.0000  1.0000  0.0000  1.0000 0.0000
--------------------------------------------------------------------------------
Option B: FOM = 0.000 deg.    MONOCLINIC   P-lattice    R(sym) = 0.059 [  1574]
Cell:     8.319   5.833  21.639   90.00  101.04   90.00    Volume:      1030.60
Matrix: 0.0000  0.0000  1.0000  0.0000  1.0000  0.0000 -1.0000 0.0000 -1.0000
--------------------------------------------------------------------------------
Option C: FOM = 0.041 deg.    MONOCLINIC   C-lattice    R(sym) = 0.322 [  1735]
Cell:     8.319  42.477   5.833   90.00   90.00   90.04    Volume:      2061.20
Matrix: 0.0000  0.0000 -1.0000 -2.0000  0.0000 -1.0000  0.0000 1.0000  0.0000
--------------------------------------------------------------------------------
Option D: FOM = 0.041 deg.    MONOCLINIC   C-lattice    R(sym) = 0.347 [  1636]
Cell:    42.477   8.319   5.833   90.00   90.00   89.96    Volume:      2061.20
Matrix:-2.0000  0.0000 -1.0000  0.0000  0.0000  1.0000  0.0000 1.0000  0.0000
```

This shows that the structure has metric orthorhombic symmetry to a good approximation. The comparison of the R_{int} values makes clear that the correct Laue group is only monoclinic, but because of the higher metric symmetry there is the possibility of twinning by pseudo-merohedry: The additional twofold axis, which is present in the orthorhombic system but not in the monoclinic one, is the twin law. To describe this axis in the monoclinic system three matrices need to be multiplied:

$$\begin{pmatrix} \text{orthorhombic} \\ \downarrow \\ \text{monoclinic} \end{pmatrix} \begin{pmatrix} \text{twofold} \\ \text{axis} \end{pmatrix} \begin{pmatrix} \text{monoclinic} \\ \downarrow \\ \text{orthorhombic} \end{pmatrix}$$

The last matrix is given by XPREP, the first one is the inverse of the last. For the twofold axis care must be taken that the monoclinic symmetry axis is not used. As a result of the transformation to orthorhombic, b is no longer the monoclinic axis, so we do not have standard monoclinic setting. We have the following matrices:

$$\begin{pmatrix} -0.5 & 0.5 & 0 \\ 0 & 0 & 1 \\ 1 & 0 & 0 \end{pmatrix} \begin{pmatrix} -1 & 0 & 0 \\ 0 & 1 & 0 \\ 0 & 0 & -1 \end{pmatrix} \begin{pmatrix} 0 & 0 & 1 \\ 2 & 0 & 1 \\ 0 & 1 & 0 \end{pmatrix} = \begin{pmatrix} 1 & 0 & 1 \\ 0 & -1 & 0 \\ 0 & 0 & -1 \end{pmatrix}$$

To check if this matrix is reasonable the following tests can be performed:

1. The matrix must transform the cell into an equivalent cell, which means that all cell constants remain nearly unchanged by this transformation. This can be checked with XPREP (option U, unit cell transformation).
2. The matrix must not be a symmetry operator of the Laue group of the structure. In the above example that would mean that the monoclinic twofold axis was used, inappropriately, and the final matrix would be `-1 0 0 0 1 0 0 0 -1`.
3. The refinement of the BASF factors is reasonable (that means the value is between 0 and 1 and the standard uncertainty is relatively small).
4. The `TWIN` command must improve the refinement. If the `BASF` refines to ~0.5 and there is no improvement, this could mean that the assumed twin axis is a crystallographic axis and the space group is wrong.

With the following two commands included into the file pmero02.res (and saved as the new .ins file pmero03.ins) the twinning is taken into account.

```
TWIN 1 0 1 0 -1 0 0 0 -1
BASF 0.2
```

The twin refinement clearly resulted in an improvement. Table 7.3 compares the two refinements with and without the TWIN command.

Although the fractional contribution of the second domain is only 7%, the refinement improves significantly. This makes clear that it is always important to check for twinning when there is higher metric symmetry. The structure had been published before (Fukuyo *et al.*, 1982), but the figures of merit were not better than ours, so perhaps this data set was also twinned, but the twinning was not detected.

Table 7.3 Comparison of the two different refinements, with and without taking the twinning into account

	without TWIN	with TWIN
$R1(F > 4\sigma(F))$	0.071	0.047
$wR2$ (all data)	0.198	0.123
k_2	—	0.073(2)
resid. electron density [eÅ^{-3}]	0.26	0.20
s.u.(C–C) [Å]	0.004 to 0.005	0.003
K (weakest reflections)	6.814	0.955

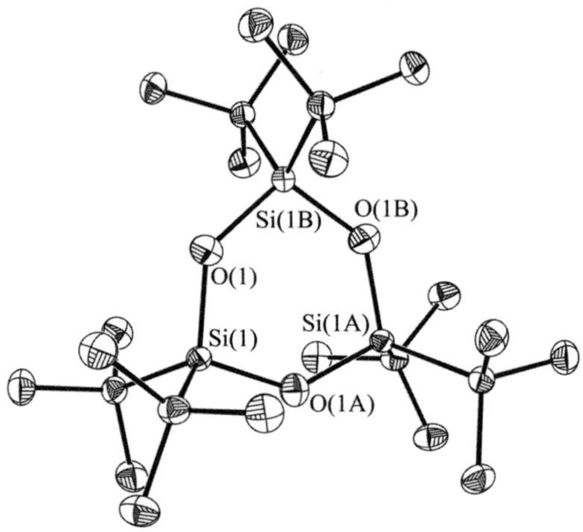

Fig. 7.11 Final model of the structure of 2,2,4,4,6,6-hexa-*t*-butylcyclotrisiloxane, corresponding to ret1-03.res. Hydrogen atoms have been omitted for clarity.

7.8.3 First example of twinning by reticular merohedry

The first example is the structure of 2,2,4,4,6,6-hexa-*t*-butylcyclotrisiloxane (Herbst-Irmer and Sheldrick, 2002). The space group determination with XPREP gives the following result (see ret1.prp) (Figure 7.11):

```
SPACE GROUP DETERMINATION

Lattice exceptions: P      A     B     C     I      F     Obv   Rev   All

N (total) =          0 24004 23981 24079 23964 36032  31915 31944 47964
N (int>3sigma) =     0  6903  6913  7404  6931 10610   3990  6964 13592
Mean intensity =   0.0  80.3  81.4  84.3  80.8  82.0   16.8  66.2  81.0
Mean int/sigma =   0.0   4.1   4.1   4.3   4.1   4.1    1.6   3.4   4.0

Crystal system H and Lattice type O selected

Mean |E*E-1| = 0.860 [expected .968 centrosym and .736 non-centrosym]

Chiral flag NOT set

Systematic absence exceptions:

        61/65 62=31  63    -c-   --c

N         33     0   33  1559   855
N I>3s     0     0    0    31   607
<I>      3.5   0.0  3.5   5.0 491.8
<I/s>    0.7   0.0  0.7   0.7  14.0
```

```
Identical indices and Friedel opposites combined before calculating R(sym)

Option  Space Group  No.  Type   Axes  CSD  R(sym) N(eq)  Syst. Abs.   CFOM

[A] R3c           #161  non-cen  1     80  0.040   3899  0.7 /  4.0    5.75
[B] R-3c          #167  centro   1     61  0.040   3899  0.7 /  4.0    5.76
```

The crystal appears to be trigonal with $a = b = 10.0789(9)$ and $c = 48.409(4)$ Å. There seem to be systematic absences for an obverse setting, although some of the reflections with $-h+k+l = 3n$ have small but significant intensity. The systematic absences for a c glide plane are obvious. So space groups $R3c$ and $R\bar{3}c$ are possible. The mean value of $|E^2 - 1|$ lies between the value for a non-centrosymmetric and a centrosymmetric space group. Structure solution with direct methods succeeds without problems in both space groups (see the files ret1-01a.res and ret1-01b.res and the corresponding .lst files). In $R\bar{3}c$ a twofold axis through the silicon and the oxygen atoms can be identified. Therefore the higher symmetry space group $R\bar{3}c$ is the correct one with one sixth of a molecule in the asymmetric unit. The refinement is straightforward but the R-values refine to only moderately low values (see the files ret1-02.res and ret1-02.lst and Table 7.4) and the highest residual electron density, which cannot be interpreted as disorder or additional solvent, is a symptom that something may be wrong. In addition there is an alarmingly long list—almost 3000—of systematic absence violations.

A closer look into the .lst file can help: all the 'most disagreeable reflections' are observed much stronger than calculated ($F_o^2 \gg F_c^2$) and for all of them $l = 3n$:

```
Most Disagreeable Reflections (* if suppressed or used for Rfree)

  h   k   l       Fo^2        Fc^2   Delta(F^2)/esd Fc/Fc(max) Resolution(A)

 -4   5   6    2533.63      558.62        8.69         0.058        1.85
 -6   9   3     895.02       18.13        7.41         0.010        1.10
 -3   3  36     826.00       34.79        5.93         0.014        1.22
  0   3  48    2116.82      844.95        5.00         0.071        0.95
  0   3  18   13667.78     8467.65        4.89         0.225        1.97
 -3   3   6   32147.31    20892.20        4.86         0.353        2.74
 -1   5  15     924.14      373.79        3.75         0.047        1.64
 -6   9   9     397.89       61.52        3.41         0.019        1.08
 -2  10   3     923.69      404.23        3.16         0.049        0.95
 -1   5  18    1171.89      643.93        2.92         0.062        1.55
  0   4   2    7687.22     5629.73        2.90         0.183        2.17
 -3   9   6     995.41      510.83        2.85         0.055        1.09
 -5   7  30     275.01       38.03        2.62         0.015        1.06
 -1   2  15   16473.84    12870.77        2.60         0.277        2.72
 -3   9  15    1866.63     1226.55        2.46         0.085        1.04
  0   3  12    4061.67     2949.40        2.46         0.133        2.36
 -2  10   6    2145.21     1461.83        2.37         0.093        0.95
  0   6  42    7760.51     6013.58        2.32         0.189        0.90
```

Table 7.4 Comparison of the two different refinements of 2,2,4,4,6,6-hexa-t-butylcyclotrisiloxane, with and without taking into account the twinning

	without TWIN	with TWIN
$R1(F > 4\sigma(F))$	0.058	0.035
$wR2$ (all data)	0.164	0.090
k_2	—	0.151(4)
resid. electron density [eÅ^{-3}]	0.95	0.40
$s.u.$(C–C) [Å]	0.004–0.005	0.002–0.003
K (weakest reflections)	4.567	2.500

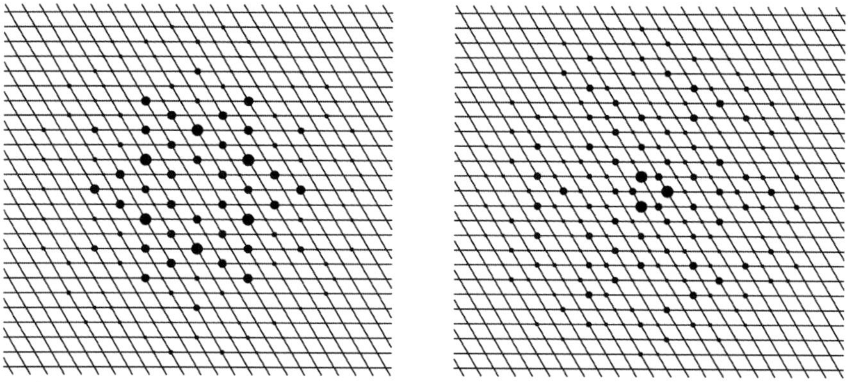

Fig. 7.12 Two reciprocal space plots (projection along l with h down and k to the right) for 2,2,4,4,6,6-hexa-t-butylcyclotrisiloxane. Left: Layer $l = 0$; right: Layer $l = 2$.

This can be explained by obverse/reverse twinning. Only the reflections with $l = 3n$ have a contribution from the second domain, and thus a measured intensity higher than one would calculate for the model. Closer inspection of the systematic absences for the lattice centring (see above) and reciprocal space plots (see Figure 7.12) confirm this hypothesis: in all layers with $l = 3n$, the reflections with $-h+k+l \neq 3n$ are absent. So only one third of the reflections are observed. For the layers with $l \neq 3n$, one third are absent, only the reflections with $-h + k + l \neq 3n$ and with $h - k + l \neq 3n$.

The obverse/reverse check in XPREP (see the file ret1.prp) gives the same result and estimates the fractional contribution of the second domain as 0.16:

```
Obverse/reverse test for trigonal/hexagonal lattice
Mean I:  obv only  145.5, rev only  28.0, neither obv nor rev 4.8
Preparing dataset for refinement with BASF 0.161 and TWIN -1 0 0 0 -1 0 0 0 1
Reflections absent for both components will be removed
```

To refine the obverse/reverse twin, edit the file ret1-02.res: you have to add `BASF 0.16` (the twin ratio suggested by XPREP) and to change `HKLF 4` to `HKLF 5`. The twin refinement uses an HKLF5-format file, derived as described above. To generate this file, the reflections are first merged according to $R\bar{3}c$ symmetry, and then the HKLF 5-format file is produced (ret1-03.hkl). This is done following the explanation given above and usually requires the use of a program which you will have to write yourself. The following is an excerpt of the file ret1-03.hkl to show what the format looks like.

```
...
   1  -8   0     2.34     2.46  -2
  -1   8   0     2.34     2.46   1
   3  -9   0     7.71     2.73  -2
  -3   9   0     7.71     2.73   1
   0  -9   0    42.70     5.06  -2
   0   9   0    42.70     5.06   1
   5 -10   0    75.67     5.18  -2
  -5  10   0    75.67     5.18   1
   2 -10   0    38.81     3.57  -2
  -2  10   0    38.81     3.57   1
   4 -11   0    76.65     4.30  -2
  -4  11   0    76.65     4.30   1
  -1   1   1     1.59     0.79   1
   0   2   1     1.83     1.04   1
  -2   3   1   920.59     4.95   1
  -4   4   1     1.29     1.56   1
...
```

All reflections with $-h+k+l \neq 3n$ do not have a contribution from the main domain and are therefore omitted. All reflections with $-h+k+l = 3n$ and $h-k+l \neq 3n$ have only a contribution from the main domain and are therefore assigned the batch number 1. Reflections with $-h+k+l = 3n$ and $h-k+l = 3n$ have contributions from both domains, and are therefore split into their two components with the indices $-h$, $-k$, l and h, k, l and batch numbers -2 and 1. The absolute value of the batch number indicates the domain number, while the combination of $-$ and $+$ tells the program that these two reflections contribute to the same measured intensity.

Refinement as an obverse/reverse twin leads to a significant improvement (see Table 7.4 and the files ret1-03.res and ret1-03.lst); Figure 7.11 shows the final model. The structure was previously determined from an untwinned crystal (Clegg, 1982) and the quality of the twin refinement is comparable to that of the original untwinned refinement.

7.8.4 Second example of twinning by reticular merohedry

The structure determination of the next example, an AlLiF cage, is not as straightforward (Hatop *et al.*, 2001; Herbst-Irmer and Sheldrick, 2002). XPREP does not identify the lattice centring and proposes a primitive lattice (see ret2.prp):

```
Original cell in Angstroms and degrees:

   14.899    14.899    30.472     90.00     90.00    120.00

  124456 Reflections read from file ret2.hkl; mean (I/sigma) =   8.47


Lattice exceptions: P     A     B     C     I     F   Obv   Rev    All

N (total) =         0 62289 62289 62272 62291 93425 82924 82920 124456
N (int>3sigma) =    0 24836 24918 24978 24949 37366 16134 21440  49852
Mean intensity =  0.0   5.5   5.6   5.5   5.5   5.5   1.9   3.5    5.5
Mean int/sigma =  0.0   8.6   8.6   8.6   8.6   8.6   3.5   5.8    8.6
```

However, comparing the mean intensity of all reflections (5.5) with the mean intensity of the reflections that should be absent in the obverse setting (1.9) and reverse setting (3.5), respectively, and inspection of reciprocal space plots (Figure 7.13) give a first indication of obverse/reverse twinning.

Additionally the high value for $\langle |E^2 - 1| \rangle = 1.068$ shows that there are a lot of weak or unobserved reflections. XPREP confirms the twinning with its obverse/reverse twin test:

```
Obverse/reverse test for trigonal/hexagonal lattice
Mean I:  obv only      9.7,   rev only      5.0,   neither obv nor rev      0.1
Preparing dataset for refinement with BASF 0.342 and TWIN -1 0 0 0 -1 0 0 0 1
Reflections absent for both components will be removed
```

The composition of the compound was not known, but an $AlC(SiMe_3)_3$ unit and some fluorines were expected. Direct methods in *R*3 with the original data (files ret2-01.ins and ret2-01.hkl) give rise to a solution that shows this moiety, but the

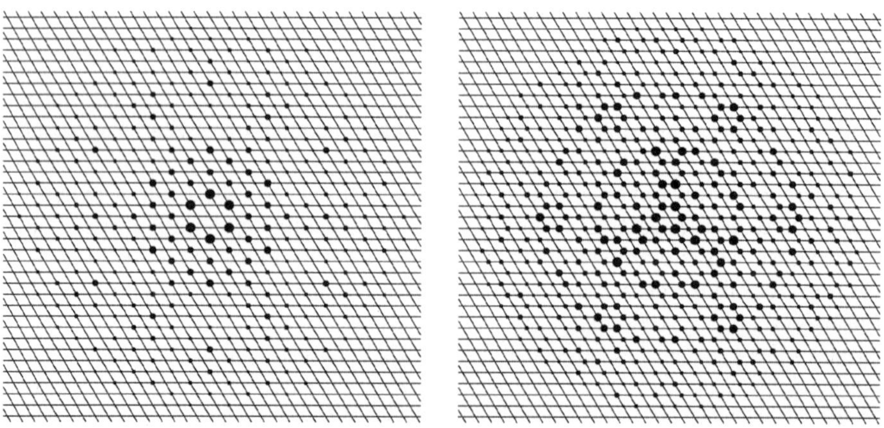

Fig. 7.13 Two reciprocal space plots (projection along *l* with *h* down and *k* to the right) for the AlLiF cage structure. Left: Layer $l = 0$; right: Layer $l = 1$.

$C(SiMe_3)_3$ group is disordered and the entire model does not appear to be very clear (files ret2-01.res and ret2-01.lst). Using SHELXL with a HKLF5 format (ret2-02.hkl) file for obverse/reverse twinning to expand from this unit leads after several steps to the whole structure. As in the example above you need to include a `BASF` into the new .ins file (the twin ratio suggested by XPREP was 0.34, that makes a good starting value) and change `HKLF 4` to `HKLF 5`. This 'twin reflection file' was derived as described above and has the following format:

```
...
   8 -16   0     4.04    0.16  -2
  -8  16   0     4.04    0.16   1
   5 -16   0     4.30    0.15  -2
  -5  16   0     4.30    0.15   1
   2 -16   0     0.98    0.16  -2
  -2  16   0     0.98    0.16   1
  10 -17   0     1.86    0.12  -2
 -10  17   0     1.86    0.12   1
   7 -17   0     2.06    0.14  -2
  -7  17   0     2.06    0.14   1
  -1   1   1   110.03    0.68   1
  -3   2   1   114.59    0.49   1
   0   2   1    27.38    0.35   1
  -5   3   1    51.24    0.29   1
  -2   3   1    45.62    0.35   1
   1   3   1    29.53    0.32   1
  -7   4   1    16.40    0.17   1
  -4   4   1    10.97    0.20   1
  -1   4   1    16.35    0.24   1
   2   4   1   122.91    0.46   1
  -9   5   1    11.53    0.16   1
```

After the first refinement, using all tricks for refining disorder as described in Chapter 5, some more carbon atoms, another fluorine, a Li(thf) and a LiO unit can be found and written into the file ret2-03.ins. At this point we can already refine the Al and Si atoms anisotropically. In the next step, (see ret2-03.res) all missing carbons of the $SiMe_3$-groups and the second thf, which is disordered about the three-fold axis (use `PART -1`), can be added (see ret2-04.ins). Then hydrogen atoms can be included and all non-hydrogen atoms are refined anisotropically (ret2-05.res). Around the three-fold axis, residual electron density is found, which can be interpreted as a disordered tetrahydrofuran molecule, refined as C_5-ring in ret2-06.res. Anisotropic refinement and adding hydrogens to this group improves the refinement only a little (ret2-07.res).

Although the whole structure (see Figure 7.14) can be found and the disorder can be modelled, the refinement remains unsatisfactory. Even though it is much better than a refinement with the original data without taking the twinning into account (see Table 7.5), we should try to find something else to further improve our structure.

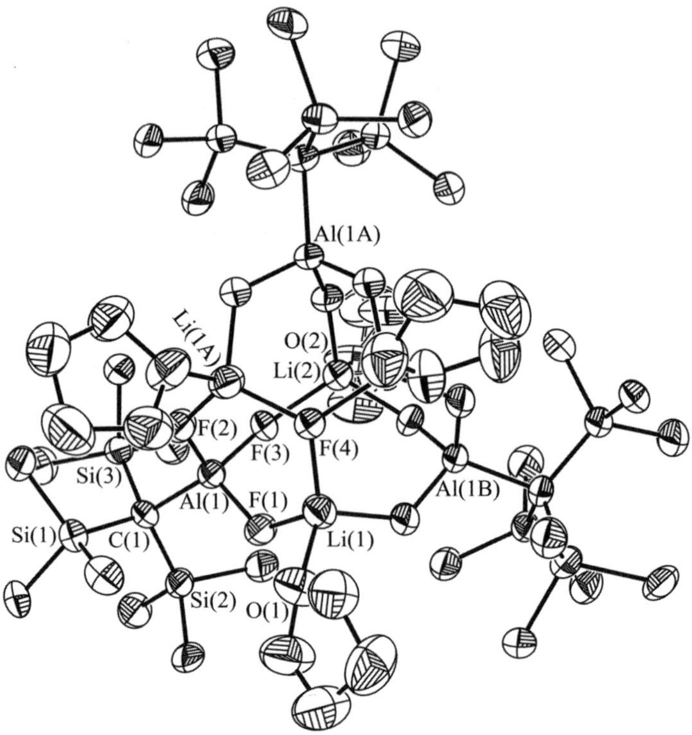

Fig. 7.14 Final model of the crystal structure of the AlLiF cage compound (corresponding to the file ret2-10.res). The disordered free thf molecule, the minor component of the disorder and hydrogen atoms have been omitted for clarity.

Table 7.5 Comparison of the three different refinements of the AlLiF cage, without taking the twinning into account, with only obverse/reverse twinning and with additional merohedral twinning taken into account

	Without twinning: ret2-08.lst	Oberse/reverse twin: ret2-07.lst	Additional merohedral twinning: ret2-09.lst
$R1(F > 4\sigma(F))$	0.149	0.112	0.034
$wR2$ (all data)	0.419	0.335	0.093
k_2	—	0.219(5)	0.004(2)
k_3	—	—	0.135(1)
k_4	—	—	0.339(2)
resid. el. density [eÅ^{-3}]	1.25	0.65	0.28
s.u.(Al–F) [Å]	0.006–0.007	0.005–0.006	0.002
Flack x	0.4(6)	0.3(5)	0.3(2)

For comparison we should perform one refinement against the original data. To do this edit the file ret2-07.res: remove the `BASF` parameter, change `HKLF 5` to `HKLF 4` and save the file as ret2-08.ins. A refinement using the original HKLF4 format file gives rise to the files ret2-08.res and ret2-08.lst. In this .lst file the $l = 3n$ reflections do not show up as clearly in the list of 'most disagreeable' reflections as they did for the first example in this section (ret1-02.lst). Further inspection of the data with XPREP using only the data fulfilling the obverse setting shows warning signs of additional merohedral twinning (see ret2.prp):

```
Crystal system H and Lattice type O selected

Mean |E*E-1| = 0.669 [expected .968 centrosym and .736 non-centrosym]

Chiral flag NOT set

Systematic absence exceptions:

        61/65 62=31  63    -c-   --c

N          33     0    33  2691  1450
N I>3s     33     0    33  1961  1383
<I>     251.1   0.0 251.1  19.8  43.5
<I/s>   114.1   0.0 114.1  22.9  40.8

Identical indices and Friedel opposites combined before calculating R(sym)

Option  Space Group  No.  Type    Axes  CSD  R(sym) N(eq)  Syst. Abs.   CFOM

[A] R-3              #148  centro    1   232  0.020   4397  0.0 / 15.0  10.05
[B] R3               #146  chiral    1    85  0.020   4397  0.0 / 15.0   2.28
[C] R3m              #160  non-cen   1    39  0.237   5361  0.0 / 15.0   9.22
[D] R32              #155  chiral    1    29  0.237   5361  0.0 / 15.0  10.05
[E] R-3m             #166  centro    1    28  0.237   5361  0.0 / 15.0  18.67
```

The mean $|E^2 - 1|$ value is 0.669 and, lower than the expected value of 0.736 for a non-centrosymmetric space group. The R_{int} value for the higher symmetry space group $R\bar{3}m$ is 0.237 compared to 0.020 for the correct space group. The difference between 0.020 and 0.237 clearly indicates the correct space group to be $R\bar{3}$ but 0.237 is small enough for additional twinning with the twin law $0\ 1\ 0\ 1\ 0\ 0\ 0\ 0\ -1$. This becomes much clearer with a second data set (ret2a.hkl). This data set was integrated on an R lattice. No signs of obverse/reverse twinning are noticed at first (see ret2a.prp):

```
SPACE GROUP DETERMINATION

Lattice exceptions: P      A      B      C      I      F     Obv    Rev   All

N (total) =         0   3628   3634   3660   3636   5461     0   4865  7279
N (int>3sigma) =    0   3558   3562   3594   3563   5357     0   4731  7121
Mean intensity =  0.0   40.9   40.4   40.5   40.6   40.6   0.0   39.5  40.6
```

```
Mean int/sigma =    0.0    53.8    53.5    53.7    53.6    53.7     0.0    51.9   53.5

Crystal system H and Lattice type O selected

Mean |E*E-1| = 0.587 [expected .968 centrosym and .736 non-centrosym]

Chiral flag NOT set

Systematic absence exceptions:

        61/65 62=31   63     -c-     --c

N           4     0     4    480     243
N I>3s      4     0     4    441     241
<I>     199.2   0.0 199.2   53.2    75.5
<I/s>   210.9   0.0 210.9   58.1    91.5

Identical indices and Friedel opposites combined before calculating R(sym)

Option  Space Group  No.   Type   Axes   CSD  R(sym) N(eq)   Syst. Abs.    CFOM

[A] R-3             #148  centro    1    232  0.037   3012  0.0 / 53.5   15.12
[B] R3              #146  chiral    1     85  0.037   3012  0.0 / 53.5    3.57
[C] R3m             #160  non-cen   1     39  0.070   3966  0.0 / 53.5   14.56
[D] R32             #155  chiral    1     29  0.070   3966  0.0 / 53.5   15.40
[E] R-3m            #166  centro    1     28  0.070   3966  0.0 / 53.5   27.80

Option [B] chosen

--------------------------------------------------------------------------------

 Obverse/reverse test for trigonal/hexagonal lattice
 Mean I:  obv only    40.5,   rev only      0.0,   neither obv nor rev      0.0

 Comparing true/apparent Laue groups.  0.05 < BASF < 0.45 indicates partial
 merohedral twinning.  BASF ca. 0.5 and a low <|E^2-1|> (0.968[C] or 0.736[NC]
 are normal) suggests perfect merohedral twinning.  For a twin, R(int) should
 be low for the true Laue group and low/medium for the apparent Laue group.

 [1] -3 / -3m1: R(int) 0.042(5081)/0.069(954), <|E^2-1|> 0.571/0.568
 TWIN   0 1 0   1 0 0   0 0 -1         BASF  0.451 [C]   or   0.437 [NC]
```

The mean value for $|E^2 - 1|$ is 0.587, even lower than in the first data set, and the R_{int} values are 0.037 for space group $R\bar{3}$ and 0.070 for $R\bar{3}m$. Significantly different R_{int} values for the higher symmetry Laue group with different crystals of the same compound clearly shows that the lower symmetry Laue group is correct and indicates different extents of twinning.

For the first data set now four twin domains should be taken into account. Reflections that are only present for the obverse setting are split into the two components $kh - l$ and hkl with batch numbers -3 and 1, while reflections with $l = 3n$ are split into the four components: $-k - h - l$, $kh - l$, $-h - kl$, and hkl with the batch numbers assigned as −4, −3, −2, and 1. So the HKLF5-format file (ret2-09.hkl) looks like this:

```
   7  -6  13     3.49     0.14  -3
  -6   7 -13     3.49     0.14   1
   7  -3  13    46.84     0.24  -3
  -3   7 -13    46.84     0.24   1
   7   0  13    15.82     0.17  -3
   0   7 -13    15.82     0.17   1
...
  -5  -2  12    37.71     0.24  -4
   5   2  12    37.71     0.24  -3
  -2  -5 -12    37.71     0.24  -2
   2   5 -12    37.71     0.24   1
  -5  -5  12    30.62     0.21  -4
   5   5  12    30.62     0.21  -3
  -5  -5 -12    30.62     0.21  -2
   5   5 -12    30.62     0.21   1
  -6   9  12    61.33     0.28  -4
   6  -9  12    61.33     0.28  -3
   9  -6 -12    61.33     0.28  -2
  -9   6 -12    61.33     0.28   1
  -6   6  12    37.90     0.25  -4
   6  -6  12    37.90     0.25  -3
   6  -6 -12    37.90     0.25  -2
  -6   6 -12    37.90     0.25   1
  -6   3  12    13.65     0.16  -4
   6  -3  12    13.65     0.16  -3
   3  -6 -12    13.65     0.16  -2
  -3   6 -12    13.65     0.16   1
  -6   0  12    45.13     0.31  -4
   6   0  12    45.13     0.31  -3
   0  -6 -12    45.13     0.31  -2
   0   6 -12    45.13     0.31   1
  -6  -3  12    68.78     0.27  -4
   6   3  12    68.78     0.27  -3
  -3  -6 -12    68.78     0.27  -2
   3   6 -12    68.78     0.27   1
...
```

To refine against this file edit the file ret2-07.ins: include two more BASF values and save as ret2-09.ins. After some ten cycles with SHELXL all refinement residuals converge to satisfactory values. After refining the weighting scheme to convergence, we have reached the publishable model (see Table 7.5 and the files ret2-09.res and ret2-09.lst).

The only remaining unsatisfactory feature is the value of the Flack x parameter. It is not possible to determine the absolute structure with certainty. We also tried the feature in SHELXL for introducing additional racemic twinning, so that we could be sure that the correct absolute structure of each domain is used, but this did not improve matters. Of course Al and Si do not have large anomalous signals with Mo radiation. But for untwinned data sets, better standard uncertainties for the Flack *x* parameter would be expected in such cases. One concludes that, for doubly twinned data sets, large anomalous signals would be needed to determine the absolute structure.

For this compound two data sets were collected. With the second data set (not shown) a refinement taking only the merohedral twinning into account (that means with a single `TWIN` command) also leads to satisfactory results. No significant sign of obverse/reverse twinning was apparent for this second crystal, but taking additional obverse/reverse twinning into account in the refinement leads to small but significant improvements, although the fractional contribution is only 9%. The hint that obverse/reverse twinning might be present came only from the first data set, which is discussed here. This suggests that obverse/reverse twinning with a small amount of the second domain may be overlooked very easily, especially if the data are integrated on an *R*-lattice. Therefore, the possibility of obverse/reverse twinning should be checked much more often or even routinely for every structure in a rhombohedral space group.

7.8.5 *First example of non-merohedral twinning*

The next structure is methylene diphosphonic acid, $CH_6O_6P_2$ (DeLaMatter *et al.*, 1973; Peterson *et al.*, 1977; Herbst-Irmer and Sheldrick, 1998) (see Figure 7.15). The data were collected years ago on a four-circle diffractometer with scintillation counter. The space group is unambiguously determined as $P2_1/c$ (see nonm1.prp) and direct methods solve the structure without problems (nonm1-01.res and nonm1-01.lst). Even though the refinement leading to the model represented by the files nonm1-02.res and nonm1-02.lst proceeds without major problems, the final *R*-values are too high (see Table 7.6) for such a simple structure. A closer look into the .lst file reveals that there are several reflections that violate the systematic absences. For most of them $|h|$ is either 6 or 1:

```
 h   k  l   Fo^2     Sigma   Why rejected

-6   0  1  930.25    15.73  Observed but should be systematically absent
 6   0  1  161.36     7.76  Observed but should be systematically absent
-6   0  3   82.29     6.27  Observed but should be systematically absent
-1   0  3  285.20     5.90  Observed but should be systematically absent
 6   0  3  130.25     7.62  Observed but should be systematically absent
-6   0  5  398.92    10.99  Observed but should be systematically absent
-1   0  5  259.21     6.96  Observed but should be systematically absent
-6   0  7  293.22    10.05  Observed but should be systematically absent
-4   0  7   19.65     4.08  Observed but should be systematically absent
...
```

In addition, many reflections disagree substantially with the model. For all of these reflections $|h|$ is 0, 1, 5, or 6, and for all of them F_o is greater than F_c:

```
Most Disagreeable Reflections (* if suppressed or used for Rfree)

  h   k   l       Fo^2       Fc^2   Delta(F^2)/esd  Fc/Fc(max) Resolution(A)

 -6   2   6     599.54       0.07        7.27         0.002        1.12
  1   2   2     589.98      10.93        6.96         0.029        2.34
  1   1   3    1942.05     641.36        6.51         0.223        2.94
 -1   2  10     596.64      33.61        6.41         0.051        1.22
  1   3   4    1497.31     438.10        6.38         0.184        1.53
  1   2   4     597.96      42.32        6.32         0.057        1.96
 -1   5   1     504.69      16.78        6.18         0.036        1.08
 -1   2   6    1444.97     352.30        5.96         0.165        1.75
 -6   3   2     698.88     111.98        5.72         0.093        1.06
  1   2   3     797.58     158.84        5.66         0.111        2.15
 -5   1   6     750.53     140.20        5.65         0.104        1.37
  1   4   3     547.64      62.99        5.52         0.070        1.27
 -5   2   1    1037.63     308.63        5.30         0.155        1.35
 -5   1  10     629.57     114.66        5.22         0.094        1.12
 -5   3   5    1490.13     609.48        4.96         0.217        1.15
  0   1   4   15509.19   11272.04        4.91         0.934        2.84
  6   1   3     583.27     124.66        4.73         0.098        1.12
 -5   4   1     418.72      59.88        4.60         0.068        1.03
  1   2   8     485.63      89.53        4.56         0.083        1.35
  5   0   4     517.55     105.91        4.50         0.091        1.27
 -1   3   6     759.30     247.84        4.27         0.139        1.42
 ...
```

Furthermore, the factor K in the variance analysis is very high for the reflections with low intensity:

```
Analysis of variance for reflections employed in refinement
K=Mean[Fo^2] / Mean[Fc^2]   for group
Fc/Fc(max)     0.000    0.016    0.032    0.049    0.065    0.086  0.111   ...
Number in group     170.     164.     177.     160.     167.     178.  ...
            GooF     1.250    1.269    1.072    1.160    0.999    1.348    ...
            K       11.456    3.036    1.540    1.398    1.174    1.210    ...
```

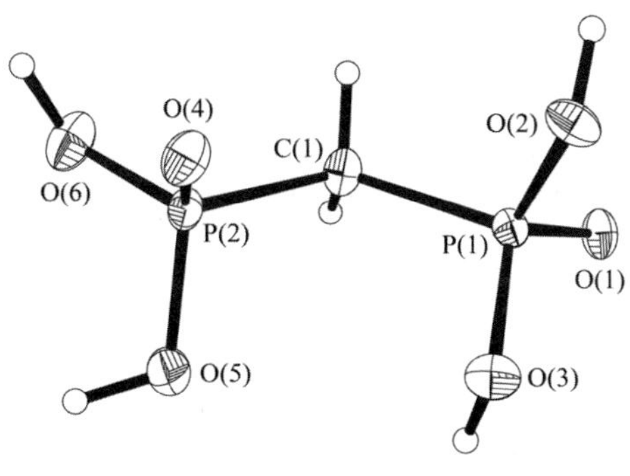

Fig. 7.15 Final model of the structure of methylene diphosphonic acid, corresponding to nonm1-07.res.

Table 7.6 Comparison of the different refinements of methylene diphosphonic acid

	Original data (no twinning) nonm1-02.lst	Omit worst reflections nonm1-03.lst	Omit $\lvert h\rvert = 0, 1, 5, 6$ nonm1-04.lst	Split $\lvert h\rvert = 0$ or 6 nonm1-05.lst	Split $\lvert h\rvert = 0, 1, 5, 6$ nonm1-06.lst	Omit worst reflections nonm1-07.lst
unique data[1]	1667	1651	991	1683	1699	1691
$R1(F > 4\sigma(F))$	0.106	0.094	0.040	0.083	0.044	0.039
$wR2$ (all data)	0.310	0.260	0.095	0.260	0.169	0.106
K^2	11.456	7.945	−0.223	4.281	0.318	0.365
resid. density [eÅ^{-3}]	2.01	1.79	0.28	1.38	0.60	0.43
k_2	—	—	—	0.241(9)	0.211(3)	0.233(2)
s.u.(P-O, P-C) [Å]	0.007–0.009	0.005–0.007	0.003–0.004	0.006–0.008	0.003	0.002–0.003

[1]The higher number of unique data in the refinements nonm1-05 to nonm1-07 compared to the original data are caused by reflections that are absent for the major domain but have intensity for the minor domain.
[2]$K = \text{mean}(F_o^2)/\text{mean}(F_c^2)$ for $0 < F_c/F_{c_{max}} < 0.016$.

There are also many high residual electron density peaks with values of more than 1 e$^-$/Å^3. Omission of the 'most disagreeable reflections' with $\Delta F^2/s.u. > 6$ (files nonm1-03.*)[10] lowers the R-values and also the residual density, but the result is not yet satisfactory. Disorder or solvent molecules are not detectable.

Again interpretation of the twinning solves the problem. To derive the twin law, we had to find a matrix that transforms the cell into an equivalent cell. The program ROTAX easily finds the following twin law:

```
Twofold rotation about   1.  0.  0. direct lattice direction:
 1.000  0.000  0.000
 0.000 -1.000  0.000
-0.822  0.000 -1.000

Figure of merit =      0.1011
```

As 0.822 ≈ 5/6 the reciprocal lattices coincide nearly exactly when $|h| = 0$ or 6. When $|h| = 1$ or 5 the reflections are so close that in most of the cases they cannot be resolved from one another. In an attempt to address this, all reflections with $|h| = 0, 1, 5$, or 6 have been omitted from the file nonm1-04.hkl; apart from the `TITL` line and the changed weighting scheme, the file nonm1-04.ins is identical to the file nonm1-02.ins. The R-values drop substantially and the residual density is now in the normal range. The K-value [mean(F_o^2)/mean(F_c^2)] for the reflections with $0 < F_c/F_{c_{max}} < 0.016$ is now very low. The most disagreeable reflection has ΔF^2/s.u. = 6.86.

Rather than omitting overlapping reflections we should take twinning for the reflections with $|h| = 0$ or 6 into account, using an HKLF5 format file with these reflections split into their two components (nonm1-05.hkl).[11] Compared to the refinement against the original data (refinement nonm1-02), the R-values drop, but there were still many reflections with $|h| = 1$ or 5, which are inconsistent. Therefore we also split the reflections with $|h| = 1$ and 5, which gives rise to the HKLF5 format file nonm1-06.hkl. The file nonm1-06.ins is identical with nonm1-05.ins (only difference in the title and the weighting scheme). This refinement is better, but now, as the .lst file reveals, the intensities of some reflections with $|h| = 1$ are underestimated:

```
Most Disagreeable Reflections (* if suppressed)

   h   k   l        Fo^2      Fc^2   Delta(F^2)/esd  Fc/Fc(max)  Resolution(A)

   1   0  -4      -42.15   2191.35       12.63          0.447          3.35
   1   0  -8       -7.32    414.93       11.57          0.195          1.71
  -1   1   3      -10.93    236.29       10.12          0.147          3.38
  -4   2   8       83.02      9.46        7.01          0.029          1.27
   1   0  -6       -8.10     39.53        6.70          0.060          2.28
 -10   1   6       88.03     15.08        6.55          0.037          0.77
```

[10] This is done by editing the file nonm1-02.ins, including an '`OMIT` *h k l*' for the top entries of the table of 'Most Disagreeable Reflections' found in the file nonm1-02.lst and saving it as nonm1-03.ins.

[11] To generate the input file nonm1-05.ins, you can start with the file nonm1-04.res, include a BASF parameter (say `BASF 0.2`) and change the `HKLF 4` to `HKLF 5`.

```
 -3   0   4     181.05      85.40     5.37      0.088      2.29
 10   1   0      48.31       8.64     5.08      0.028      0.75
  1   4   1       0.51      24.05     4.44      0.047      1.33
-10   0  10     116.45      46.87     4.00      0.065      0.74
 -7   1  13      26.15       2.53     3.39      0.015      0.84
  7   0   2     434.74     284.63     3.38      0.161      1.03
 -1   0   6     228.86     147.56     3.34      0.116      2.28
  7   2   7      65.47      30.12     3.24      0.052      0.82
-10   2   9     123.59      71.55     3.16      0.081      0.72
-10   2  10      24.18       0.16     2.93      0.004      0.71
  1   0 -12      -3.01      13.38     2.84      0.035      1.14
```

For these reflections we only have partial overlap, while the refinement assumes exact overlap. Therefore, in an approach similar to that used in nonm1-03.ins, we again omit the worst reflections from this list, say all reflections with ΔF^2/s.u. $>$ 6. Edit the file nonm1-06.res accordingly and save as nonm1-07.ins.

After refining the weighting scheme to convergence we have arrived at our final model. This structure is known and in the literature there is no indication of twinning. Our final refinement results in similar standard uncertainties to the published untwinned structure, although our *R* value is higher, probably because of the problem in handling partial overlap; we also suspect that one twin component was better centred in the beam than the other. Nowadays, with area detectors and suitable programs for handling non-merohedral twins, the procedure is much more convenient as will be shown with the following example.

7.8.6 Second example of non-merohedral twinning

The data for the structure of 2-(chloro-methyl)pyridinium chloride (Jones *et al.*, 2002) was collected on a Bruker SMART 1000 CCD area detector. The normal indexing program failed and split reflections profiles combined with nice profiles and reflections very close to each other indicated that this was a non-merohedral twin. The program CELL_NOW easily finds two orientation matrices (see nonm2·_cn):

```
-------------------------------------------------------------------------------

Cell for domain  1:    7.433    7.869   12.607    89.17    89.24    78.01

 Figure of merit: 0.753   %(0.1):  79.5   %(0.2):  82.9   %(0.3):  85.9

 Orientation matrix: -0.00192477  0.09460331  0.05237783
                     -0.00459288  0.08608949 -0.05954688
                     -0.13745303  0.02278947  0.00209452

 Percentages of reflections in this domain not consistent with lattice types:
 A: 45.2,  B: 51.8,  C: 51.0,  I: 45.4,  F: 74.0,  O: 67.1  and  R: 68.8%

 Percentages of reflections in this domain that do not have:
 h=2n: 44.1,  k=2n: 49.0,  l=2n: 49.9,  h=3n: 69.0,  k=3n: 69.7,  l=3n: 71.0%
```

```
  465 reflections within 0.200 of an integer index assigned to domain  1,
  465 of them exclusively;    96 reflections not yet assigned to a domain

-----------------------------------------------------------------------------

Cell for domain  2:    7.433    7.869   12.607    89.17    89.24    78.01

Figure of merit: 0.705   %(0.1):  81.3   %(0.2):  97.9   %(0.3):  99.0

Orientation matrix: -0.03941386  0.09480299 -0.05366872
                    -0.03459287  0.08567318  0.05840024
                     0.12715352  0.02351695 -0.00165375

Rotated from first domain by 179.5 degrees about
reciprocal axis -0.003  1.000  0.004  and real axis -0.223  1.000 -0.006

Twin law to convert hkl from first to      -0.999  -0.005   0.006
this domain (SHELXL TWIN matrix):          -0.445   0.999  -0.011
                                           -0.017   0.006  -1.000

RLATT color-coding employed in file: 6ad2.p4p
White: indexed for first domain
Green: current domain (but not in a previous domain)
Red: not yet indexed

  304 reflections within 0.200 of an integer index assigned to domain  2,
   94 of them exclusively;     2 reflections not yet assigned to a domain

-----------------------------------------------------------------------------
```

The twin law is a twofold rotation about the axis 0 1 0, which is common. SAINT V7.12a is used for the integration with both orientation matrices, giving rise to the file nonm2-m.mul and, among other information, the following output (nonm2-m._ls):

```
Statistics for reflections in nonm2-m.mul
File is in BrukerAXS area detector ASCII format
Histograms will be accumulated for component  1

Spots with multiple components (twin overlaps) will not be included in
histograms

Number of spots read from file =            25520
Number of components read from file =       32160
Number of component 1 singlets =             9501
Number of component 2 singlets =             9379

Number of spots with 1 component =          18880
Number of spots with 2 components =          6640 (excluded from histograms)
Number of spots with 3 components =             0 (excluded from histograms)
Number of spots with 4 components =             0 (excluded from histograms)
```

```
Number of spots with invalid component number =        0
Number of spots with < 1 component =            0
Number of spots with > 4 components =           0

Occurrences of overlaps between components:

                        and        2        3        4
Between component
       1                        6640        0        0
       2                                    0        0
       3                                             0
```

There are 25520 spots, 9501 of them have only a contribution from the first domain and 9379 only from domain 2; 6640 spots belong to both components. TWINABS V1.02 is used for scaling and absorption correction (see nonm2.abs):

```
  9501 data (    3350 unique ) involve component  1 only,   mean I/sigma  20.3
  9379 data (    3287 unique ) involve component  2 only,   mean I/sigma   7.7
  6640 data (    2798 unique ) involve   2 components,   mean I/sigma  20.4
```

The first domain is much bigger than the second, as can be seen from the mean I/sigma for the two domains. For the structure solution a detwinned data file in HKLF4 format is prepared in TWINABS. XPREP easily determines the space group $P2_1/c$ (see file nonm2.prp):

```
SPACE GROUP DETERMINATION
Lattice exceptions:  P     A     B     C     I     F    Obv   Rev   All

N (total) =          0  2288  2290  2264  2266  3421  3039  3048  4564
N (int>3sigma) =     0  1657  1713  1650  1671  2510  2243  2247  3382
Mean intensity =   0.0  13.1  15.7  16.3  16.8  15.0  16.1  15.7  16.1
Mean int/sigma =   0.0  22.6  24.7  25.2  24.5  24.2  25.1  24.7  24.8

Crystal system M and Lattice type P selected

Mean |E*E-1| = 0.924 [expected .968 centrosym and .736 non-centrosym]

Chiral flag NOT set

Systematic absence exceptions:

        -21-   -a-   -c-   -n-

N          9    89    90    93
N I>3s     0     2    41    41
<I>      0.1   0.1  24.1  23.3
<I/s>    0.4   1.0  20.4  19.5

Identical indices and Friedel opposites combined before calculating R(sym)

Option  Space Group  No.  Type   Axes  CSD  R(sym) N(eq)  Syst. Abs.   CFOM

[A] P2(1)/c         # 14  centro   4 19410  0.019   2164  1.0 / 19.5   0.74
```

The structure is solved without any problems (nonm2-01.res and nonm2-01.lst).

For the refinement TWINABS can also produce an HKLF5 format file, but XPREP had transformed the cell constants to derive a standard setting for space group $P2_1/c$. As an HKLF5 format file cannot be read by XPREP and a matrix on the HKLF card is not possible with `HKLF 5`, the cell constants for the integration must be in standard setting prior to the generation of the HKLF5 format file. In CELL_NOW it is possible to transform cell constants, and a new .p4p file containing the two orientation matrices in the correct setting can be generated (see nonm2b._cn). Then the integration must be repeated with the cell in the standard setting (giving rise to nonm2b-m.mul and nonm2b-m._ls) before TWINABS produces the HKLF5-format file. All reflections with a contribution from the main domain are used. All overlapped reflections are split into their two components with batch numbers -2 and 1, respectively, and all non-overlapped reflections have the batch number 1 (file nonm2-02.hkl):

```
...
  -1  18   1     0.55     4.69   1
   0  18   1    83.51     5.89   1
   1  18   1     0.68     4.69   1
  10   0   2    12.45     7.51  -2
 -11   0   2    12.45     7.51   1
   9   0   2   267.16    10.74  -2
 -10   0   2   267.16    10.74   1
   8   0   2   686.97     7.14  -2
  -9   0   2   686.97     7.14   1
   7   0   2   451.40     4.40  -2
  -8   0   2   451.40     4.40   1
   6   0   2   797.27     4.38  -2
  -7   0   2   797.27     4.38   1
   5   0   2   197.26     2.08  -2
  -6   0   2   197.26     2.08   1
...
```

The structure refines to good results: $R1(F > 4\sigma(F)) = 0.027$, $wR2(\text{all data}) = 0.071$, residual density $= 0.43\text{e}\text{Å}^{-3}$, see Figure 7.16).

However, the number of unique reflections is artificially high: 3435, compared to 2302 reflections for the detwinned file. In the list of 'most disagreeable reflections' in the file nonm2-02.lst, most of them still have $F_\text{o} > F_\text{c}$.

```
Most Disagreeable Reflections (* if suppressed or used for Rfree)

    h   k   l       Fo^2          Fc^2   Delta(F^2)/esd   Fc/Fc(max)  Resolution(A)

    3   2   2     146.96         97.17        7.55          0.084          1.85
    3   0   2      37.58         18.55        6.60          0.037          1.94
    0   6   0     117.33        163.90        6.06          0.110          2.14
   -2   0   4      85.39         59.82        5.96          0.066          1.82
   -5  14   4     344.83        255.06        5.85          0.137          0.76
    2   3   9      43.69          9.25        5.58          0.026          0.76
```

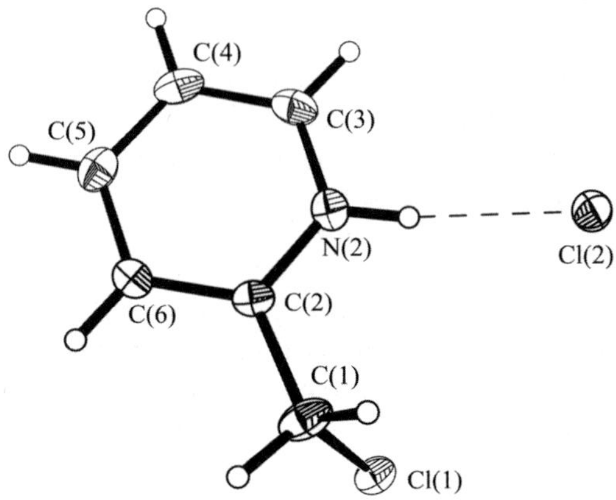

Fig. 7.16 Final model of the structure of 2-(chloro-methyl)pyridinium chloride corresponding to nonm2-03.res.

```
   4   5   2     135.33       96.62      5.09     0.084     1.35
  -6   0  10     120.98      175.31      5.03     0.113     0.71
   3   4   0      51.18       70.26      4.97     0.072     2.02
  -8   5   7     321.20      247.69      4.74     0.135     0.77
  -2   5   2     108.00       85.61      4.14     0.079     1.95
   3   4   2     275.40      225.61      4.13     0.129     1.66
  -1   7   4     122.68       95.82      3.99     0.084     1.31
...
```

Checking the HKLF5-file reveals that for most of these, there are two entries, one without any overlap with the second domain and one with an overlap:

```
  3   2   2  103.51    1.93   1
 -4   2   2  222.75    1.73  -2
  3   2   2  222.75    1.73   1

 -4   0   2  121.97    2.08  -2
  3   0   2  121.97    2.08   1
  3   0   2   26.47    1.51   1

  3  14   4  242.87    4.98  -2
 -5  14   4  242.87    4.98   1
 -5  14   4  153.79    6.63   1
  0   0   4   58.04    3.90  -2
 -2   0   4   58.04    3.90   1
 -2   0   4   60.14    1.18   1

  4   5   2   95.32    3.41   1
 -5   5   2  210.82    2.08  -2
  4   5   2  210.82    2.08   1
```

On one frame SAINT determines those reflections (or symmetry related ones) as split and on another frame as unsplit. Of course TWINABS cannot merge an overlapped with a non-overlapped reflection. One could think of conditions that leave one reflection unaffected by the twinning but affect a symmetry-related reflection, but the number of such reflections should be small. However, in our tests more than 20% of the data have these so-called 'twin pairing errors'. Therefore we omit all reflections that are unsplit and have a split symmetry equivalent in the file nonm2-03.hkl. Although nearly 30% of the data are affected the difference in the two refinements are not significant: $R1(F > 4\sigma(F)) = 0.027$, $wR2(\text{all data}) = 0.071$, residual density $= 0.43\text{e}\text{Å}^{-3}$, see nonm2-03.lst). The number of data is now 2293. So it seems to be possible to ignore these 'twin pairing errors' in routine structure determinations, assuming that the redundancy is high enough.

7.9 Conclusions

Twinning usually arises for good structural reasons. When the heavy atom positions correspond to a higher symmetry space group it may be difficult or impossible to distinguish between twinning and disorder of the light atoms (Hoenle and von Schnering, 1988). Since refinement as a twin usually requires only two extra instructions and one extra parameter, in such cases it should be attempted first, before investing many hours in a detailed interpretation of the 'disorder'! Refinement of twinned crystals often requires the full arsenal of constraints and restraints, since the refinements tend to be less stable, and the effective data to parameter ratio may well be low. In the last analysis, chemical and crystallographic intuition may be required to distinguish between the various twinned and disordered models, and it is not easy to be sure that all possible interpretations of the data have been considered.

8

Artefacts

'Artefact'—this word is one of the most overused and abused terms in crystallography. Some crystallographers tend to explain most problems they cannot solve with either 'packing effects' or 'artefacts'. Like every physical method, crystallography is affected by errors. However, there are two kinds of errors: unavoidable errors such as artefacts, and avoidable errors. A good crystallographer knows how to avoid avoidable errors, how to live with unavoidable errors and, most importantly, how to distinguish between the one and the other.

This chapter is more about scientific work ethics than about structure refinement. Nothing can be done to overcome an artefact, but the words 'this is an artefact' should never be used as an excuse for sloppy workmanship or as an explanation for effects that seem to be inexplicable. Not every crystal structure can be perfect, and there will always be cases that show some unexplained features. Creating a pseudo-explanation by calling these effects 'artefacts' holds the risk of drawing other scientists' attention away from them. Of course that is precisely the reason for using the term 'artefact' in such a case: who would want the journal's referees to zero in on and to ask questions about something you cannot explain either? However, if this tactic should succeed, it will be even more difficult to find the correct explanation later, because nobody will think about it again. If there is an unsolved problem, should it not be the duty of a good scientist to draw the attention of the scientific community to it, and have other scientists help to find the solution? Of course, that may not be the way to be successful with the upcoming grant application, but, fortunately, this book is not about tactics to get government money.

8.1 What is an Artefact?

Webster's Encyclopedic Unabridged Dictionary of the English Language gives six definitions for the term artefact. Definition five says an artefact is 'a spurious observation or result arising from preparatory or investigative procedures', and definition six reads 'any feature that is not naturally present but is a product of an extrinsic agent, method or the like'. The word artefact is used in several contexts and disciplines, but in the world of crystallography, an artefact is a method-immanent unavoidable systematic error leading to incorrect observations. Typical crystallographic artefacts are

1. Bond lengths appearing too short due to libration.
2. Carbon–carbon or carbon–nitrogen triple bonds appearing too short due to high electron density between the nuclei.

3. Inaccurate determination of hydrogen atom positions.
4. Spurious electron density maxima (or minima), especially on or near special positions and close to heavy atoms, arising from Fourier series truncation errors (e.g. when strong reflections are missing, or for low-resolution data).

8.1.1 Libration

Atoms in a crystal are not as motionless as the physical appearance of a crystal may make them appear to be. In fact, as we saw in Chapter 5, some atoms can actually move quite strongly in a crystal. Even in molecules without obvious disorder, atoms are never totally still nor do they exhibit displacement evenly in all directions, which is the reason why anisotropically refined structures give rise to much better models. As mentioned before, collecting data at low temperature drastically reduces molecular and atomic motion, giving rise to better diffraction data, but even at 0 K, atoms would show zero-point vibration. Besides temperature, the location and mass of an atom influence its displacement: terminal atoms are free to move much more strongly than atoms in the core of molecules, and, if a relatively light atom is bound to a heavier one, the lighter atom, having less inertia, will show the stronger motion of the two. This form of motion, an oscillation mostly perpendicular to the direction of the bond between the atoms, is called libration. Yet how can atomic motion influence the bond length determined by X-ray diffraction?

Figure 8.1 shows a simple example with two covalently bound atoms, A and B at distance r from one another, where one atom (atom B) shows much stronger libration than the other one. When we describe the spatial distribution of the atomic positions in the form of ellipsoids, which is commonly done in anisotropically refined crystal structures, the average position of an atom is calculated as the centre of the ellipsoid. For atoms that show strong libration, the centre of the ellipsoid can be significantly far away from the true position of the atom, which leads interatomic distances to appear shortened by Δr. As libration is more pronounced at higher temperature, this effect is more noticeable in room temperature structures than in structures determined at low temperature (say 100 K). This is the reason why interatomic distances determined with X-ray diffraction tend to be shorter at higher temperature, even though this seems to contradict common physical experience. This effect is due to libration and is an artefact of the method of modelling structures.[1] As described, it affects mainly light atoms, which are terminally bound and can be in the range of up to 0.2 Å.

A simple equation estimates the value of Δr, which can be used to perform a libration correction. In the case of room-temperature structures it can be appropriate to calculate libration-corrected distances for terminal atoms:

$$\Delta r \approx \frac{\Delta U}{2r} = \frac{[U_{\mathrm{B}} - U_{\mathrm{A}}]}{2r} \tag{8.1}$$

[1] There are alternative approaches to describe the electron density function of an atom more accurately than with a simple ellipsoid, e.g. as banana-shaped ellipsoids or in form of electronic multipoles (Bader, 1990).

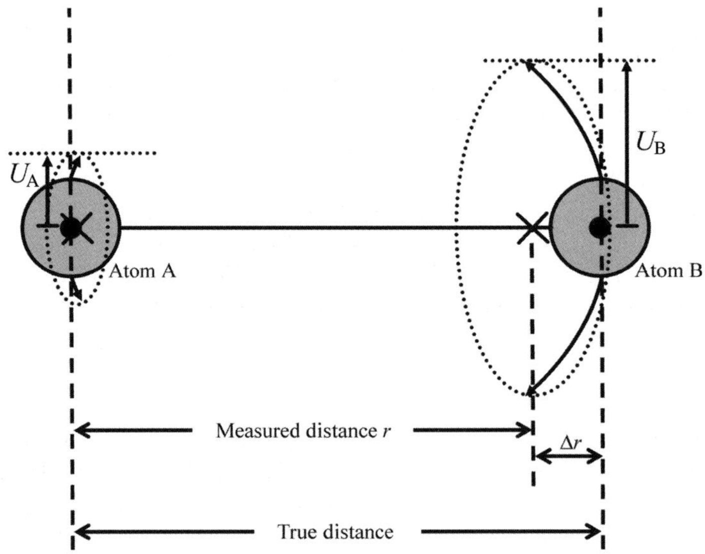

Fig. 8.1 Apparent bond-shortening due to libration in the example of two atoms, drawn as grey circles. The atomic displacement of atom A is described by U_A, that of atom B by U_B. The measured distance r is too short by the length Δr. The ellipsoids fitting the atomic motion are drawn as dotted lines; the centre of each ellipsoid, which corresponds to the apparent atomic location, is marked with a large X. The true distance is between the centres of the atoms, marked as black dots.

The above equation does not take into account the movement of other atoms and assumes that the libration described by x_A and x_B is only due to vibration of the atoms relative to one another and is not a result of displacement of the two atoms together as a rigid group. A better but more complicated description of libration and how to correct for it is given in Schomaker and Trueblood (1968).

For some high temperature structures that show strong libration, it can be appropriate to calculate and publish libration corrected distances. One possible way to perform this correction is the `LIBR` command in XP.

8.1.2 Shortened triple bonds

It is the electrons in a crystal that interact with the X-ray photons, and the electron density distribution is the quantity actually measured in crystallography. When taking the step from the electron density map to the atomic model, the simple assumption is that locations of high electron density correspond to atomic positions. Albeit a very simple and maybe a little naïve assumption, the models derived with its help are mostly very sensible and accurate. There are cases, however, when this assumption does not hold.

Triple bonds between light atoms, that is carbon–carbon or carbon–nitrogen triple bonds, show relatively high electron density between the atoms, when compared to the number of electrons located at the sites of the nuclei. Calculating the three-dimensional coordinates of two electron density maxima describing the two atoms involved in such a triple bond will place them too close together, thus making the bond appear too short (assuming the true bond length is the distance between the nuclei). As this effect is unavoidable and method-immanent, the shortening of triple bonds between light atoms is an artefact.

As long as atoms are not described as electronic multipoles (Bader, 1990), there is no way to overcome or to correct for this effect. It is important to always bear in mind that bond lengths determined by X-ray diffraction are distances between electron density maxima and not between the true positions of the nuclei.

8.1.3 Hydrogen positions

For reasons very similar to those just described, it is challenging to locate hydrogen atoms in an X-ray crystal structure: there is only one electron and its distribution is not anywhere close to the location of the proton. In addition, hydrogen is not only the lightest element, but binds mostly terminally to other atoms. Therefore hydrogen is affected by libration much more than any other element. This effect adds to the difficulties of determining the exact hydrogen positions by X-ray diffraction.[2] The positioning and treatment of hydrogen atoms in X-ray crystal structures is of such high importance for structure determination that there is an entire chapter on hydrogen atoms in this book (Chapter 3) and the inaccuracies of hydrogen positions as well as the short X-H bond lengths can be seen as crystallographic artefacts.

8.1.4 Fourier truncation errors

The electron density function is given in the form of a Fourier summation (Equation 4.2). This means that electron density and hence the atoms in a structure are represented by a number of sine-waves, which are added up. The higher the number of sine-waves, the smoother and more accurate the electron density becomes. As with every Fourier summation, if terms are missing, ripples appear. Especially when some strong reflections are missing from the dataset (e.g. incomplete dataset or some reflections hidden behind the beamstop) artefactual electron density—negative or positive—can appear near heavy atom sites. The same effect can be observed with low-resolution data. An excellent description of the theory behind this effect can be found in an article by Cochran and Lipson (1966).

A famous example is the structure of Nitrogenase MoFe-Protein, a protein that contains a Fe_7MoS_9 cluster. The inside of this cluster is about 4 Å wide with six iron atoms closest to the centre, and older crystal structures had been determined at resolutions of about 2 Å. Termination of the Fourier summation at that resolution creates an artefactual minimum in the electron density of about −0.2 electrons about 2 Å away from each iron atom. These spurious minima from all heavy atoms in the

[2] When it is vital to determine the hydrogen positions with high accuracy, neutrons should be used instead of X-rays in the diffraction experiment.

cluster overlap in the centre of the cage, almost perfectly hiding a nitrogen atom that can be found with better data. In this example, artefactual negative electron density kept scientists from finding the nitrogen atom in the cage. The discovery of this atom by Oliver Einsle *et al.* (2002) changed the view of the mechanism of biological nitrogen fixation.

8.2 What is not an artefact?

Besides artefacts, there are other systematic errors that can have negative effects on a crystal structure; global pseudo-symmetry for example, or inaccurate scaling. And then there is the group of avoidable errors. Common avoidable errors are:

(1) Wrong unit cell dimensions
(2) Twinned structure refined as disorder
(3) Wrong atom type assignments
(4) Wrong space group
(5) Fourier termination peaks mistaken for hydrogen atoms (or vice versa)

The German crystallographer Roland Boese (1999) made an interesting point in distinguishing between 'avoidable errors' and 'really avoidable errors'. As examples for the latter kind he listed:

(1) Typographical errors in the unit cell dimensions
(2) Misadjustment of the diffractometer (wrong zero points, etc.)
(3) Wrong data collection and/or data reduction strategy
(4) Mistakes in the refinement
(5) Data collection at room temperature

Those 'really avoidable errors' are surprisingly common and can have a multitude of negative effects. In the last few years, software, mostly developed by diffractometer manufacturers, has become more and more fool-proof. As with almost everything new, this is both good and bad. Good, because many mistakes can be avoided with the help of these programs, and bad, because inexperienced crystallographers get away with insufficient knowledge too easily.

8.3 Example

The following example of a refinement shows some spurious electron density maxima, which can be understood as an artefact.

8.3.1 Fourier termination error in $C_{30}H_{47}N_9Zr_5$

The Zr(IV) compound $C_{30}H_{47}N_9Zr_5$ crystallizes in the form of green prisms in the tetragonal space group *I*4. The asymmetric unit contains a quarter of the molecule, and the rest is generated by the crystallographic fourfold axis. The core of the structure consists of five Zr atoms forming a square pyramid. The four triangular faces of this pyramid are capped by NH groups; the four edges of the base are bridged by NH_2 groups, and in the centre of the basal plane of the Zr_5 cluster there is a μ_5-N atom. The coordination sphere of the Zr atoms is completed by one methyl-cyclopentadienyl (MeC_5H_4) group per metal (see also Bai *et al.*, 2000).

The files zr5-00.ins and zr5-00.hkl on the accompanying CD-ROM are the SHELXS input and reflection files for this example. The solution, zr5-00.res, contains three independent Zr positions: Zr(1) on a general position, Zr(2) and Zr(3) on the crystallographic fourfold. This arrangement gives rise to a Zr_6 tetrahedron, when the crystallographic fourfold is applied. The meaning of the other electron density maxima is not entirely clear at this stage, thus all hypothetical Q-atoms are deleted, and only the three independent Zr atoms are transferred into the file zr5-01.ins.

After 15 cycles of refinement[3] with SHELXL, one independent MeC_5H_4 ligand, coordinated to Zr(1), becomes visible. It is formed by residual electron density maxima Q(6), Q(7), Q(8), Q(9), Q(11), and Q(12). The U_{eq} of one of the Zr atoms on the fourfold, Zr(3), has refined to a very large value. Therefore, Zr(3) is deleted and only Zr(1) and Zr(2) as well as the MeC_5H_4 ligand are kept and saved as zr5-02.ins.

After six cycles of refinement (zr5-02.res), the following assignments can be made. Q(1) is in the centre of the pyramid and is probably a nitrogen atom (say N(3)). Q(2) and Q(5) are too close to Zr(2) to be atoms and can be ignored for the moment, as they will probably disappear once the atom is refined anisotropically. Electron density maximum Q(3) and its three symmetry equivalents correspond to nitrogen atoms capping the four triangular faces of the pyramid, let's call it N(2). Peak Q(4) and its symmetry equivalents correspond to nitrogen atoms bridging the four edges of the pyramid's base (rename Q(4) as N(1)). Q(6) and its three symmetry equivalents form a square where the remaining MeC_5H_4 ligand would be expected (see Figure 8.2). This is a typical case in which part of a molecule does not fulfil the space group symmetry. This ligand has to be refined using `PART -1`, as described in Chapter 5. Before this can be done, however, we need to find the fifth carbon atom or somehow come up with it. This can be achieved using the constraint `AFIX 56`, which makes the five atoms following the command form a perfect pentagon. To do this, generate the three symmetry equivalents of Q(6) (e.g. using the `grow` command in XP) and assign them to be carbon atoms C(21) to C(24). The fifth carbon atom, C(25) we are going to 'make up', assigning it the coordinates (0, 0, 0). The five atoms are preceded by `PART -1` and `AFIX 56` and followed by `AFIX 0` and `PART 0`:

```
PART -1
AFIX 56
C21    1   -0.01390    0.08000   -0.41010   10.250000   0.05000
C22    1    0.08000    0.01390   -0.41010   10.250000   0.05000
C23    1    0.01390   -0.08000   -0.41010   10.250000   0.05000
C24    1   -0.08000   -0.01390   -0.41010   10.250000   0.05000
C25    1    0          0           0         10.250      0.05
AFIX 0
PART 0
```

[3] The choice of 15 cycles is somewhat arbitrary, but 10 seemed to be insufficient. See also Chapter 12.

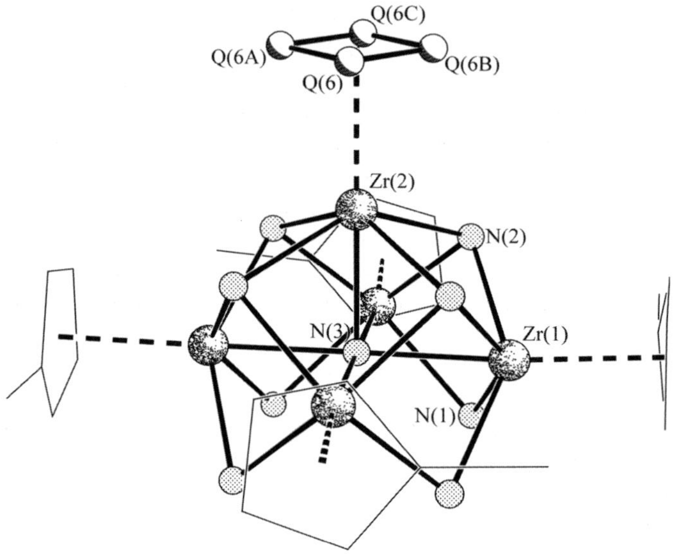

Fig. 8.2 Interpretation of Q(1) as N(3), Q(3) as N(2) and Q(4) as N(1) in the file zr5-02.res. Q(6) and its symmetry equivalents correspond to a Cp-ring, which is disordered about the crystallographic fourfold.

In addition to these changes, we should stabilize the position of the five-membered ring relative to Zr(2) by restraining all Zr–C distances to be equivalent using

```
SADI zr2 c21 zr2 c22 zr2 c23 zr2 c24 zr2 c25
```

and allow the two independent Zr atoms to be anisotropic by adding `ANIS $zr`. The file zr5-03.ins contains all these changes.

Six cycles of refinement with SHELXL give rise to much better *R* values and a message from the program that our absolute structure is probably wrong. A look into the file zr5-03.lst tells us that the Flack-*x*-parameter (Flack, 1983) has refined to 0.9(2). We will therefore invert the structure before the next cycle. This can be done in XP (using the `INVT` command) or with the help of the `MOVE` command in SHELXL: The coordinates of atoms that follow `MOVE dx dy dz sign` are shifted from their original position by `dx`, `dy`, `dz` if `sign` is 1. If `sign` is −1 the atoms are shifted and inverted (see the SHELXL manual for further details). Thus, for most space groups, inversion of the structure can be achieved with the line `MOVE 1 1 1 -1` in the .ins file before the first atom, which is also appropriate here.[4]

[4] For the eleven space groups that come in enantiomorphous pairs (like $P3_1$ and $P3_2$), the translation parts of the symmetry operators need to be inverted as well to generate the other member of the pair, In addition there are seven space groups where inversion on the origin does not lead to the inverted absolute structure (see Bernadinelli and Flack, 1985). Here is a list of those seven space groups and the corresponding correct `MOVE` instructions:

*Fdd*2	`MOVE 0.25 0.25 1 -1`				$I4_1cd$	`MOVE 1 0.5 1 -1`				
$I4_1$	`MOVE 1 0.5 1 -1`				$I\bar{4}2d$	`MOVE 1 0.5 0.25 -1`				
$I4_122$	`MOVE 1 0.5 0.25 -1`				$F4_132$	`MOVE 0.25 0.25 0.25 -1`				
$I4_1md$	`MOVE 1 0.5 1 -1`									

The residual electron density maximum Q(2) corresponds to the missing methyl group, bound to C(24) and is named C(20) (do not forget to change the occupancy to 0.25!). In addition, the `AFIX 56` / `AFIX 0` constraint can be replaced with similarity restraints to make the two crystallographically independent MeC_5H_4 ligands equivalent: the lines `SAME c10 > c15` and `SAME c20 c21 c25 < c22` are added right before C(20) in the file zr5-04.ins. In order for this to work, the carbon atoms in the second ring need to be renamed, as the methyl group is supposed to bind to C(21). If you do not rename the carbon atoms, the `SAME` restraint will treat atoms as equivalent, which, in fact, are not. In order to avoid fluctuations in the U_{eq} values of the carbon atoms, `SIMU` and `DELU` can be added for all atoms (see Section 2.6.2).

In the file zr5-04.res, a second position of the first MeC_5H_4 ligand, the one not affected by the fourfold axis, becomes visible: residual electron density maxima Q(8), Q(5), Q(6), Q(3), Q(2), and Q(7) correspond to atoms C(10A) to C(15A). Using our refinement skills from Chapter 5, we can formulate this disorder, as it has been done in the file zr5-05.ins.

The refined model in zr5-05.res is a lot more convincing than before. Now we can try to refine all atoms anisotropically by adding `ANIS` in the file zr5-06.ins.

This results in two carbon atoms being non-positive definite. A quick (and hopefully temporary) fix is to make the `SIMU` and `DELU` stricter for the carbon atoms. Change the two commands in the file zr5-06.ins and save as zr5-06a.ins.

This time all atoms refine well, and we find residual electron density maxima representing the three independent hydrogen atoms on the nitrogen atoms N(1) (Q(15) and Q(17)) and N(2) (Q(3)) in the difference Fourier synthesis. These three hydrogen atoms have been included into the file zr5-07.ins (don't forget the `DFIX` commands for the N—H distances). We should also include the appropriate `HFIX` commands to include the hydrogen atoms that bind to carbon.

The file zr5-07.res represents the complete anisotropic model with all hydrogen atoms. Now, the weighting scheme can be adjusted and refined to convergence. The problems with the carbon atoms, especially of the second ligand, remain. Apparently, the combination of the crystallographic fourfold with the fivefold symmetry of the Cp-ligand causes correlations among parameters of the atoms of the second ligand, which cannot be overcome by `PART -1` alone. The relatively strict restraints on the displacement parameters (`SIMU` and `DELU`) can be set back to default values, however introduction of equal displacement parameter constraints (`EADP`) for all atoms of the second ligand are necessary. All this has been done in the file zr5-08.ins.

When we analyze the list of residual electron density peaks in the file zr5-08.res, the final model of this refinement, we find that Q(1) and Q(2) are significantly higher than all the other maxima. Q(1) corresponds to 1.38 electrons and is located 0.66 Å away from Zr(1), not far from the location of the deepest hole. This could be an artefact arising from Fourier truncation, but absorption effects can also lead to spurious electron density close to heavy atom positions. It is difficult to decide which of the two effects holds responsibility for Q(1), but it is clear that we cannot do anything about it. Q(2) represents 1.13 electrons and sits on the crystallographic fourfold. It

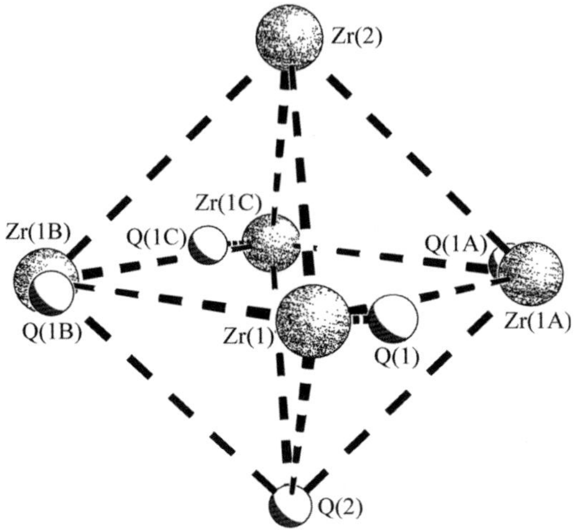

Fig. 8.3 Zr-atoms of the final model of $C_{30}H_{47}N_9Zr_5$, with the two highest residual electron density maxima (all carbon, nitrogen and hydrogen atoms have been omitted for clarity). Q(2) is located on a special position, completing un-observed symmetry.

takes the position of Zr(3) in the original solution from SHELXS (zr5-00.res), thus completing the Zr(5) square pyramid to an octahedron. This is a typical artefact: spurious electron density on a special position, completing un-observed symmetry.[5] Figure 8.3 shows the Zr-atoms of the final model with the two highest residual electron density maxima and their symmetry equivalents.

[5] Another possible explanation for Q(2) could be a 95:5 or so disorder of Zr(2) and its ligands. Such a disorder, however, should also result in a higher U_{eq} value for Zr(2), which is not observed. Actually, the opposite is the case: $U_{eq}(Zr(1)) = 0.030$, $U_{eq}(Zr(2)) = 0.024$. This difference can be explained with the special position constraints (see Section 2.5.2) that restrict the shape of the displacement ellipsoid of Zr(2) to fulfil the fourfold symmetry. This, in turn, artificially lowers the calculated value of U_{eq} for Zr(2).

9

Structure validation

Anthony L. Spek

The impetus of chemists to carry out an X-ray structure determination is clearly worded by Dorothy Hodgkin in her 1964 Nobel Lecture: '[The] great advantage of X-ray analysis as a method of chemical structure analysis is its power to show some totally unexpected and surprising structure with, at the same time, complete certainty.' This point is clearly demonstrated in the fields of organometallic and coordination chemistry with the large number of supporting crystal structures that appear in journals such as *Inorganic Chemistry* and *Organometallics*. Unfortunately, the literature is also full of 'surprising results' that are seriously wrong. They are invariably based on sloppy experimentation and inadequate analysis. Often, papers include discussions of 'interesting features' of a structure that in hindsight turn out to be based on artifacts. The erroneous concept of 'bond-stretch isomerism' (Parkin, 1993) is a good example.

This chapter will point out some of the more common and avoidable pitfalls. Readily available computer software will be discussed that generates appropriate ALERTS when something unusual or inconsistent is detected. The analyst should, of course, understand the meaning of a particular ALERT and be able to respond with proper action.

A crystal structure determination is not finished with the refinement to convergence of the structural model parameters such as the positional and displacement parameters. Incorrect models may well result in satisfactory looking convergence of the least-squares refinement. The model has to be analyzed in detail in a process called structure validation. In particular, the model should make chemical sense. An incorrect model may easily lead to disastrous conclusions about the underlying chemistry. Examples can be found in Harlow (1996) and Spek (2003).

Before the early 1990s structure validation was largely based on the experience of the crystallographic investigator and that of the referees of the associated scientific publications. Since then the number of experienced investigators and knowledgeable referees has decreased drastically and the number of structural studies has increased exponentially. As a consequence various errors of interpretation are easily overlooked, thus adding systematic noise to databases such as the CSD (Allen, 2002). A partial answer to this problem has been found in the form of the design and implementation of automated structure validation procedures.

The starting point of automated structure validation was the creation of a universal computer readable data format, named CIF (Hall *et al.*, 1991), for reporting the results of a crystal structure analysis. In hindsight it was a fact of extreme importance that the very popular SHELXL refinement program (Sheldrick, 1997b) was one of the early adopters of this standard that is now universal within the structural community. The next step was the use of the CIF-standard as the electronic medium for structural papers submitted for publication in *Acta Crystallographica C*. Finally, it was realized that the numerical data, now available in well-defined computer readable format, could also be used for automated inspection. The International Union of Crystallography, IUCr, has set-up a web-based tool for this purpose named checkCIF.

Serious errors of interpretation are often related to missing, too many or misplaced hydrogen atoms with big impact on the reported chemistry. Equally disastrous is the assignment of wrong atom types. Unrecognized disorder is yet another source of misinterpretation. An incorrect space group assignment may easily result in unusual geometry. Finally, data and structure quality can be seriously hampered by unresolved experimental problems such as twinning, absorption and incomplete reflection data.

9.1 Validation

Structure validation should give an answer to the following three questions:

- *Is the reported information complete*?

The answer to this question is easily handled with a checklist of items to be reported in order to be able to repeat the study. Consistency is also easily checked. Examples are the comparison of the volume as calculated from the cell dimensions with the reported volume and the density calculated from the cell content as derived from the coordinate list and symmetry with the reported density.

- *What is the quality of the analysis*?

This is a more difficult question. It depends on factors such as crystal quality, data collection hardware, expertise of the investigator and software used. Various qualifiers can be calculated and compared with some generally accepted standards. Examples are resolution and completeness of the data set, bond precision, convergence of the least-squares refinement, *R* factors and residual density excursions in the final difference map.

- *Is the structure correct*?

This is the most important and also most difficult question. A computer program can give only a partial answer to this. What software can do in this respect is to report on any unusual structural feature it detects. It is then up to the investigator to consider the issue raised, take proper action if necessary, or otherwise comment convincingly on it in the final report.

The IUCr has defined and documented a large number of validation tests (http://journals.iucr.org/services/cif/checking/autolist.html). They address the quality, completeness and consistency issues. In addition they include checks for proper refinement and absorption correction procedures. Anonymous crystal structure validation is made available through the web-based IUCr checkCIF facility at (www.journals.iucr.org/services/cif/checking/checkform.html).

Validation tests can also be run locally with the implementation of the programs IUCRVAL (Farrugia, 2000) and PLATON (Spek, 2006). This is particularly useful when direct access to the internet is a problem or not allowed in view of confidential research.

9.2 Validation tests implemented in PLATON

The numerous PLATON validation checks, also available through the IUCr checkCIF facility, address a large number of completeness, quality and correctness issues. A list of ALERTS is generated with an associated severity level A, B or C attached. Each ALERT type is accompanied with an explanation of the reported problem. Validation is generally carried out using the SHELXL generated (and suitably edited) CIF file and optionally together with the associated FCF reflection file. In the following paragraphs a number of tests will be discussed in some detail.

9.2.1 Missed symmetry

The assignment of the proper space group to a given structure is not always obvious at the beginning of a structure determination. Often, a preliminary structure can be obtained only in a space group with symmetry that is lower than the actual symmetry. Subsequent analysis should lead to a transfer into a description in the correct space group. Unfortunately, the latter is not always achieved as demonstrated many times by Dick Marsh and others (e.g. Marsh and Spek, 2001). The frequency of structures reported with too low symmetry is shown to be as high as 10% in certain space groups (e.g. *Cc*).

An attempt to refine a centrosymmetric structure in a non-centrosymmetric space group generally results in poor geometry due to the (near) singularity of the least-squares normal matrix. Chemically equivalent bonds may differ significantly and the displacement parameters generally make little sense in such a case. Proper action to solve the problem includes, apart from leaving out half of the atoms in the model and the addition of an inversion centre, a shift of the structure to the proper origin. A number of missed symmetry cases are clearly caused by a failure to do this properly.

The PLATON utility ADDSYM, an extended version of the powerful MISSYM algorithm (Le Page, 1987, 1988) is used to report possibly missed higher crystallographic symmetry. The software suggests a tentative more appropriate space group. Missed symmetry ALERTS generally require a detailed analysis. In many cases access to the primary reflection data is called for in order to distinguish cases of missed symmetry from frequently occurring cases of pseudo-symmetry.

9.2.2 Voids

The validation software also reports on solvent-accessible voids (van der Sluis and Spek, 1990) in the structure. Such voids might include disordered solvent that went undetected in the peak-search algorithm. A common reason for this might be that the disorder results in density ridges or faint plateaus rather than isolated peaks. Except for some framework structures, crystal structures generally collapse when they have lost solvent molecules of crystallization.

Voids are frequently located at or along symmetry elements. Solvent molecules on those sites are generally highly disordered or fill one-dimensional channels along three-, four- or sixfold axes.

Void ALERTS in combination with ALERTS on short inter-molecular contacts may point to molecules that are misplaced with respect to the symmetry elements.

Voids can potentially contain disordered charges with an impact on the valence state of the main residue.

9.2.3 Displacement ellipsoids

A displacement ellipsoid plot (see also Chapter 4 and Johnson, 1976) is an excellent validation tool for visual inspection by an expert, but is not suitable as an automatic analysis tool. Fortunately, the Hirshfeld rigid-bond test (Hirshfeld, 1976) provides an excellent numerical analogue. Both tools are indicative of a variety of problems with the structural model. The central idea of the Hirshfeld test is that the components of the anisotropic displacement parameters along the bond for two bonded atoms should have approximately the same value.[1] This will generally not be the case when incorrect atom types are assigned to density peaks. Carbon atoms might be nitrogen atoms or oxygen etc. Elongated ellipsoids are, in general, indicative of unresolved disorder (see also Figure 5.1). An attempt should be made to resolve the disorder.

Many systematic errors (including absorption, wrong wavelength and missed super-lattice symmetry) find their way into the displacement parameters, often giving rise to physically impossible non-positive definite values for the main-axis values.

9.2.4 Bond lengths and angles

The numerical values of bonds and angles are checked to fall within expected ranges based on the tentatively derived hybridization of the associated atoms. Bonds that are too short or too long may be artefacts of unresolved disorder. For an interesting example where the original authors interpreted a very long C—C bond as a 'transition state' see Kapon and Herbstein (1995). The average value and the range of the C—C bond lengths within a phenyl moiety are compared with the expected value, 1.395 Å. A significant deviation from that value may indicate incorrect cell dimensions (possibly calculated with the wrong wavelength), poor diffraction data or an incorrect refinement model.

9.2.5 Atom type assignment

The correct assignment of element types to electron-density maxima can be a serious pitfall (see Chapter 4 and Müller, 2001). Nitrogen and oxygen are often

[1] It is the Hirschfeld theorem that the DELU restraint in SHELXL is based on. For details see Chapter 2.

interchangeable in ring systems. Misinterpretation can have important consequences for the chemistry involved. For a detailed analysis of a wrong atom type assignment see Li *et al.* (2001). Also, the positions of N and C—H are often found exchanged in five and six-membered ring moieties.

9.2.6 Intermolecular contacts

Inspection of intermolecular contacts can be very informative in pointing at incorrect structures. Obviously, when atoms approach each other closer than the sum of their van der Waals radii there must be either a missed interaction, such as a hydrogen bond, or their positions are in some way in error. Bumping hydrogen atoms may indicate misplaced hydrogen atoms (e.g. two instead of one hydrogen on an sp^2 carbon) or methyl moieties fixed in an inappropriate conformation.

An interesting case (CSD-entry IDAKUT) where obviously no intermolecular contact analysis was performed is a reported structure with an S—H moiety (Celli *et al.*, 2001). This structure appears to contain a short S$\cdots$S contact of 2.04 Å being clearly a missed S—S bond. The H atom on S should be deleted from this wrongly reported dimeric structure.

9.2.7 Hydrogen bonds

As a rule with few exceptions, OH moieties are hydrogen bonded to an acceptor. Potential H-atom positions lie on a cone. Finding the correct position on this cone can be tricky when the difference electron-density map does not show a single suitable maximum. SHELXL (Sheldrick, 1997b) provides an option to find the optimal position by way of an electron-density calculation around a circle (see Chapter 3 for details). Inspection of contoured difference maps for hydrogen atom positions should be attempted in less obvious settings.

9.2.8 Connectivity

A proper CIF is assumed to contain a set of atomic coordinates that do not require the application of symmetry operations to connect them into chemically complete molecules and ions. The exception is where a molecule possesses crystallographic symmetry, but the asymmetric fragment should still be a connected set. Checks are performed to identify isolated atoms. An isolated transition metal likely points to a misinterpreted identity. In at least one case the real identity turned out to be a Br anion. Isolated hydrogen atoms might need a symmetry operation to bring them into a bonding position or their bond distances might be outside the expected range. Single bonded metal atoms are also flagged since they probably represent the assignment of an incorrect atom type.

9.2.9 Disorder

Reported disorder can be real or an artefact resulting from poor or erroneous experimental procedures. Severe disorder is frequently accompanied by a fast decay of the average intensity as a function of theta. If not, the disorder might be an artefact of a too small unit cell chosen for the data collection and associated averaged structure. Unresolved substitutional disorder (e.g. an admixture of Cl and CH_3 substituents) may give rise to unusual bond distances and displacement parameters. See Chapter 5 for the correct refinement of disorders.

9.2.10 Reflection data

Reflection data can be checked for completeness when made available to the PLATON validation software as an FCF file along with the CIF file. Inadequate data collection procedures may result in an incomplete survey of the reciprocal lattice. Data collection on a CCD or image-plate based diffractometer may require more than one scan in order to avoid a cusp of missing data.[2]

The analysis of the FCF reflection data for reflections with F_o significantly larger than F_c may point at unaccounted for (pseudo) merohedral twinning.

9.2.11 Refinement parameters

Unresolved issues may show up in the refinement parameters. In general, the value of $wR2$ should not be much larger than twice the value of $R1$. The second parameter in the expression for the weight should not be much larger than 1.0. The value of the S (goodness of fit) value when refined with SHELXL should be around 1.0. The final difference map should be essentially featureless with negative and positive excursions of the same order.[3]

9.3 When to validate

It is much easier to detect and correct for overlooked problems with a crystal structure during or immediately at the end of an analysis than during the publication process. Early on structure validation also allows the investigator to go back to the experiment, when appropriate, for the gathering of additional data. Validation software should be available on the platform where the structure is solved and refined or accessible on that platform through the web.

9.4 Concluding remarks

Single crystal X-ray crystallography has been and still is a unique, reliable and unbiased source of new chemical knowledge. Unfortunately, there is a catch. The investigator has to be knowledgeable about the various pitfalls that have to be avoided. Automated validation offers an unbiased list of potential problems and issues to be addressed. All ALERTS generated by the software should be analyzed

[2] In general, the quality of the data is much improved when more scans than absolutely necessary are performed. For details see also the comments on MoO (Multiplicity of observations) in Chapter 2 and Müller (2005).

[3] For definitions of $R1$, $wR2$ and S see equations 2.3, 2.4 and 2.5.

in detail. They can point either to unresolved problems or to potentially interesting new science. Everything of the type 'unusual' or 'new' should be scrutinized in detail and preferably supported by independent evidence. An excellent tool for this purpose is the Cambridge Crystallographic Database and associated software such as VISTA.

Nowadays, most structural results are published in non-crystallographic journals as part and in support of chemical studies. Regrettably, referees frequently receive insufficient information to judge the adequacy of the analysis—in particular when very limited crystallographic details are given, often stuffed in a footnote or a CSD reference number. Some journals seem not even to include a trained crystallographer as one of the referees because this holds-up rapid publication of important chemistry. Unfortunately also those only marginally checked structures subsequently go into the literature and databases as a 'refereed' publication.

10

Protein refinement

Thomas R. Schneider

Since the first protein structures were determined by X-ray crystallography in the 1950s and 1960s, macromolecular crystallography has gone through an impressive evolution. Proteins can be produced in large quantities and at high purity using recombinant expression systems and modern techniques of protein biochemistry. The identification of optimum conditions for the formation of well-diffracting crystals is greatly facilitated by the use of small volumes and the availability of sophisticated liquid handling techniques allowing users to screen vast numbers of conditions. The use of synchrotron radiation in conjunction with very sensitive detectors has revolutionized the field as it allows collecting diffraction data from ever-smaller crystals. The problem of radiation damage that occurs with the use of high doses of radiation has been partly alleviated by the development of techniques to maintain crystals at cryogenic temperatures. As a consequence of these developments, not only the structures of ever more complicated macromolecular systems can be determined, but also, for an increasing number of cases, atomic resolution data[1] can be collected on crystals of biological macromolecules.

At atomic resolution, the number of observables is much higher than at medium or low resolution (see Table 10.1) allowing for the refinement of models with more parameters representing a more detailed description of the structure and the flexibility of the molecule in the crystal, for example by using anisotropic displacements parameters (U^{11}, U^{22}, U^{33}, U^{12}, U^{13}, U^{23}) instead of isotropic B-values[2] that are usually used in macromolecular crystallography. In the electron density maps, the high resolution of the data is reflected by more details, such as multiple conformations and hydrogen atoms becoming visible. A further advantage of high resolution data is that, due to their sheer number (e.g. in Table 10.1, the observable-to-parameter ratio is still better than 5:1 at 1.1 Å resolution despite the use of 9 parameters per atom), the refined values of the parameters will be much more precise than at lower resolution allowing for example for the comparison of interatomic distances at a very fine level.

[1] A definition for the term 'atomic resolution data' has been given by Sheldrick (1990). A discussion of the properties of atomic resolution data in the context of macromolecular crystallography can be found in Morris and Bricogne (2003).

[2] The isotropic B-value that is commonly used in macromolecular crystallography is related to the equivalent isotropic displacement parameter U_{eq} by the equation $B_{iso} = 8\pi^2 U_{eq}$. A guide to the different nomenclatures used to describe atomic displacements in crystals can be found in Grosse-Kunstleve and Adams (2001) and Trueblood *et al.*, (1996).

Table 10.1 Observables and parameters in macromolecular crystallography

d_{min} [Å]	N_{obs}	parameters	N_{par}	obs : par
3.0	11,634	x, y, z	15,000	0.6 : 1
2.5	20,104	$x, y, z, (B_{iso})$	15,000	1.3 : 1
2.0	39,267	x, y, z, B_{iso}	20,000	2.0 : 1
1.5	93,078	x, y, z, B_{iso}	20,000	4.5 : 1
1.1	236,018	$x, y, z,$ $U^{11}, U^{22}, U^{33}, U^{12}, U^{13}, U^{23}$	45,000	5.2 : 1
0.9	430,919	$x, y, z,$ $U^{11}, U^{22}, U^{33}, U^{12}, U^{13}, U^{23}$	45,000	9.6 : 1

For each resolution d_{min}, the approximate number of observables N_{obs} and the observation to parameter ratio (obs : par) for a typical parameterization giving rise to a number of parameters N_{par} is given. The number of reflections has been calculated for a hypothetical structure containing 5000 non-hydrogen atoms and 40% bulk solvent in a triclinic unit cell. For crystal structures with a higher solvent content, the number of observables at a given resolution will be higher. In real cases, the number of observables should be increased by the number of restraints and the number of parameters should be decreased by the number of constraints (see also Chapter 2).

Different aspects of atomic resolution structures of biological macromolecules have been reviewed (e.g. Schmidt and Lamzin, 2002; Esposito *et al.*, 2002; Vrielink and Sampson, 2003). Examples of particular interest are:

- Detection of a covalent bond in an intermediate state of an enzymatic reaction (Heine *et al.*, 2001).
- Study of an enzyme-inhibitor complex at 0.66 Å resolution (Howard *et al.*, 2004).
- Titration of a histidine in the crystalline state (Berisio *et al.*, 1999).
- Characterization of structural consequences of radiation damage (Schröder Leiros *et al.*, 2001).
- Observation of structural changes during photo-isomerization (Genick *et al.*, 1998).

Conceptually, the refinement of a protein structure at atomic resolution with SHELXL can be considered as the refinement of a huge organic molecule using small molecule techniques as explained in other chapters of this book. Thus, apart from technical issues, such a refinement of a macromolecule with SHELXL should not represent a problem for an experienced small molecule crystallographer. For the macromolecular crystallographer, using the full repertoire of parameters and restraints and constraints may be more of a challenge. Therefore, this chapter approaches the refinement of protein structures with SHELXL more from a protein crystallographer's point of view.

In the following, the typical steps of the refinement of a protein structure at atomic resolution using SHELXL will be described. The functionality discussed can be used equally well for other macromolecules such as RNA or DNA. An overview of the features of SHELXL can be found in Sheldrick and Schneider (1997); for technical information, the reader is referred to the SHELX-website (www.shelx.uni-ac.gwdg.de/SHELX). Please note that SHELXL can also be used for the refinement of twinned protein structures (see Chapter 7) and at less than atomic resolution (e.g. Usón *et al.*, 1999).

10.1 Atomic resolution refinement *vs.* standard refinement

When macromolecular structures are refined at atomic resolution, a number of concepts that are commonly used in the refinement of small molecules have to be introduced. These include parameterizations for modeling the static and dynamic disorder of atoms in a crystal and for the positioning of hydrogens. Furthermore, the treatment of both ordered and bulk solvent and the determination of standard uncertainties via the inversion of the normal matrix of the refinement need to be discussed.

10.1.1 Anisotropic displacement parameters

The subject of an X-ray diffraction experiment and the subsequent analysis is not a single molecule at an instantaneous point in time, but an agglomerate of many billions of molecules (the crystal) observed over a time (the duration of the experiment) that is very long compared to the time scales of conformational changes within the molecules. Therefore the electron densities representing the molecule of interest will not indicate a sharp position for each atom, but due to the averaging in space (conformational heterogeneity) and in time (movements of atoms) correspond to a more or less complicated distribution (Dunitz *et al.*, 1988). In macromolecular crystallography at less than atomic resolution, this distribution is usually approximated by a Gaussian distribution centered on a single site parameterized by the position of an atom and its B-value. At atomic resolution, the large number of observables allows for a more detailed description by approximating the true distribution with a three-dimensional ellipsoid. This ellipsoid is characterized by six parameters (three for its orientation and three for its extents into different directions).

The introduction of anisotropic displacements parameters (or ADPs) into a refinement more than doubles the number of parameters from 4 (3 coordinates plus 1 B-value) to 9 (3 coordinates plus 6 ADPs) per atom and thus needs to be tightly monitored to avoid over-fitting the data. Typically, a drop in R_{free} of at least 1.0–1.5% should be observed upon switching from isotropic to anisotropic displacement parameters. In many cases, the inclusion of ADPs results in a dramatic improvement of the crystallographic phases and the corresponding electron density maps.

Similar to the use of stereochemical restraints to stabilize the refinement of atomic coordinates, anisotropic displacement parameters need to be restrained to keep them physically reasonable. The restraints implemented in SHELXL are described in

Chapter 2 and illustrated in Figure 2.2. Their weights are set by the `DEFS` instruction and normally do not need to be changed.

10.1.2 Multiple discrete sites

A common feature of an electron density map at atomic resolution is that two peaks of electron density, corresponding to one atom being present in two different locations, are found. This type of disorder can not only affect individual atoms but also group of atoms such as entire side chains (Figure 10.1) or groups of residues (Sevcík *et al.*, 2002).

To describe such a situation with the minimum number of parameters and to enforce a physically reasonable model, SHELXL allows constraining the occupancies of all atoms in a conformer to be the same while the sum of the occupancies of all conformers works out to 1.0. The non-bonded interactions between conformers (e.g. atoms belonging to the same conformers do not make non-bonded interactions with their equivalents in other conformers) are organized fully automatically by SHELXL once the different sites have been assigned to specific conformers (using the `PART` instruction as described in Chapter 5).

10.1.3 Hydrogens

Approximately half of the atoms in a protein structure are hydrogen atoms (Andersson and Hovmöller, 1998). However, due to the little electron density associated to them, hydrogen atoms are usually ignored in standard macromolecular refinement.

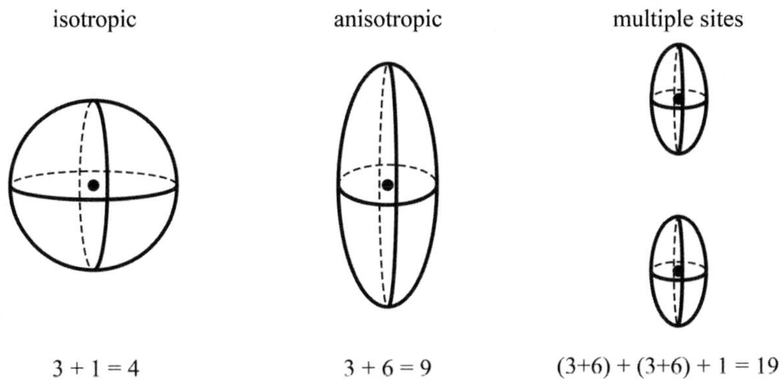

Fig. 10.1 Schematic views of different parameterizations. Isotropic refinement involves 3 parameters for the coordinates and 1 isotropic *B*-value. In anisotropic refinement, 6 parameters are used to model the probability function for the location of an atom. When an atom needs to be modeled using more than one site, the number of parameters increases accordingly. However, to define the occupancy of two sites only one extra parameter is needed, as the occupancy p_2 of the second site can be calculated from the occupancy of the first site, p_1, by $p_2 = 1.0 - p_1$.

SHELXL offers a vast number of ways to include hydrogens into a model allowing for maximum economy in terms of parameters while keeping full flexibility for all chemically possible situations (see Chapter 3 in this book). For a large portion of the hydrogens in a protein molecule, their positions will usually be calculated geometrically based on the positions of the neighbouring non-hydrogen atoms (the riding-model). For such calculated positions not a single additional parameter will enter the refinement and, in principle, there is no reason to exclude such hydrogen atoms from a refinement at less than atomic resolution.

For hydrogen atoms whose positions cannot be calculated—for example hydroxyl hydrogens, that, in the simplest approximation, are allowed to move on a circle centered on the C—O bond, thus costing one parameter—their coordinates have to be determined from the electron density or derived by inference from the surrounding hydrogen bonding pattern. However, in many cases the diffraction data are not strong enough to allow an unambiguous placement. In fact, incorrect placement of such hydrogen atoms will in many cases result in serious geometric distortions in their vicinity (Sheldrick and Schneider, 1997; Word *et al.*, 1999).

Adding hydrogen atoms to a protein model usually reduces $R1$ and $R1_{\text{free}}$ by the same amount, typically 0.5–1%. More importantly, the inclusion of hydrogens will improve the geometry of the model as the non-bonded interactions (enforced by `BUMP` restraints, see Chapter 2 and the SHELX manual for details) between atoms are described more accurately. However, the effect of the inclusion of hydrogens on the crystallographic phases and electron density maps is usually not very pronounced.

10.1.4 Solvent

At atomic resolution, not only fully occupied solvent sites, but also partially occupied solvent sites can be modeled with confidence. To facilitate a stable refinement, the occupancies of partially occupied sites should be coupled to neighbouring protein atoms and/or other partially occupied solvent sites in the neighbourhood using constraints.

10.1.5 Standard uncertainties

A large observable-to-parameter ratio allows for the estimation of standard uncertainties of the refined parameters by the inversion of the normal matrix of the refinement (Cruickshank, 1970; Press *et al.*, 2002). Due to the rapid progress in computing systems, the huge calculations involved can now be done in hours on modern desktop workstations.

Once the standard uncertainties of the parameters that were actually refined (e.g. coordinates of atoms) have been determined, the standard uncertainties of derived quantities (e.g. distances between atoms) can be calculated by error propagation. An example is the detection of the protonation of a carboxylate group. Once the positions of the C and the O atoms have been refined without restraints against atomic resolution data, the presence of a *significant* difference between the lengths of the two C—O bonds (a double *vs.* a single bond in the protonated case) can be used to detect a protonation.

10.2 Stages of a typical refinement

10.2.1 Getting started

A refinement at atomic resolution will typically be started from a mostly complete model originating, for example, from a previous structure determination of the same molecule at lower resolution or from a model created by automatic building software using an experimental or molecular replacement based phase set. Such a model will usually describe the atoms in the structure with fully occupied sites and isotropic *B*-values.

SHELXL input and output files

In contrast to most other refinement programs, SHELXL reads both the current parameters of the model and the instructions for the next round of refinement from one single file, the instruction file (or ins-file). A helper program called SHELXPRO can be used to automatically set up such an ins-file from a pdb-file. Among other items, the ins-file produced will describe all the necessary restraints for geometric parameters using target values from the Engh and Huber library (Engh and Huber, 1991) as well as restraints for anisotropic displacement parameters.

Diffraction data are kept in a fixed-format asci-file, the hkl-file. SHELXPRO can be used to convert various formats of diffraction data to an hkl-file suitable for SHELXL. More versatile for this purpose is the program XPREP (see Chapter 1); and for conversion of data from the CCP4-world, a program mtz2various is available within the CCP4 suite (Collaborative Computational Project Number 4, 1994). When converting data between formats, it is good practice to explicitly check the precise number of reflections (including the number of reflections in work and test sets used for cross-validation) whenever data are moved between different formats. Possible causes for observing inconsistent numbers of reflections can be different R_{free} flagging conventions, implicit or explicit resolution cut-offs, formatting problems for very low or high intensities, different schemes of flagging unobserved reflections, the format for negative intensities, and others. Frequently such differences will only be detected at the time of publication and then will cause considerable stress.

On Output, a refinement job will produce a res-file (which is a valid ins-file for the next round of refinement) containing the new description of the model, a pdb-file containing the coordinates of the refined model, an lst-file containing logging information, and an fcf-file containing structure factor moduli and phases for the calculation of electron density maps. The fcf-file can be read directly by Xfit (McRee, 1999) and Coot (Emsley and Cowtan, 2004) or converted into other formats with SHELXPRO.

Initial adjustment of the model

To adjust the overall scale factor, the solvent model and the position of the initial model, it may be advisable to begin a SHELXL refinement with one round of rigid body refinement. As the number of parameters is small (6 per rigid body), full matrix refinement (keyword `L.S.`) can be used. The corresponding instructions are:

```
L.S. 20 -1     ! Run 20 cycles of full matrix least
               !  squares refinement
BLOC 1         ! Refine only coordinates
SHEL 10 2.0    ! Use data between 10 and 2.0 Å
STIR 4.0 0.3 ! Begin refinement with data to 4.0 Å, then
               !  with every cycle include successive 0.3 Å
               !  shells of data
.
AFIX 6         ! constrain all atoms to be in one rigid body
.
... ATOMS ....
.
AFIX 0
```

If appropriate, the model can be partitioned into different rigid bodies, for example different chains or protein and ligands, using pairs of `AFIX 6` and `AFIX 0` instructions.

Typical problems

- When a molecular replacement solution is used as a starting model, overlapping regions of symmetry-related molecules can cause the generation of many anti-bumping restraints. These can be deactivated by out commenting the `BUMP` instruction. A more thorough solution is to first rebuild the starting model using an automated procedure such as arpWarp (Perrakis *et al.*, 1999).
- Incorrectly measured low-resolution reflections (e.g. overloaded reflections or reflections 'behind the beam-stop') can seriously impede refinement at all stages. Care must be taken to assess the correctness of these reflections. If there are any doubts, a low-resolution cut-off can be applied using the `SHEL` keyword.

10.2.2 Rough adjustments of the model at 1.5 Å

The initial refinement of scale factors is followed by a refinement of coordinates and *B*-values against data limited to 1.5 Å. This step is important to adjust the geometry of the model against the SHELXL set of restraints, as some of these have small differences with respect to restraints used in other programs. Given that the number of parameters at this stage is substantial (four parameters per atom), the conjugate gradient algorithm (`CGLS` instruction) needs to be used for refinement. This algorithm (Sheldrick and Schneider, 1997) is not as accurate as the least-squares method but reaches convergence with much less computer time. The `STIR` instruction is used to improve convergence:

```
CGLS 20 -1     ! Run 20 cycles of conjugate gradient
               !  refinement
SHEL 10 1.5    ! Use data between 10 and 1.5 Å
STIR 2.0 0.1 ! Begin refinement with data to 2.0 Å, then
               !  with every cycle include successive 0.1 Å
               !  shells of data
```

The resulting model is inspected for gross problems such as misplaced side chains and spurious water molecules. All warnings in the lst-file should be checked carefully. In particular the warnings concerning atoms without restraints and the list of 'Disagreeable restraints' can point to problematic regions and incompatibilities between stereochemical restraints used in different programs. At this stage, SHELXWAT can be used to update an existing or to build a new preliminary water structure.

Typical problems

- The refinement becomes instable when coordinates and B-values are released. In many cases, single badly defined atoms can be the culprits. Such atoms can be identified by checking the refinement parameters that display the maximum shifts as indicated in the lst-file (search for 'Max. shift' in the lst-file).
- After a transition from a different refinement program to SHELXL, the R-values are different from what you had before. This can be caused by differences in the model, for example the use of different bulk solvent corrections or by different sets of reflections used for the calculation of the R-values. In particular, the use of different sigma-cutoffs (e.g. the exclusion of reflections that have an amplitude of less than say 3 times their standard deviation, '$F < 3\sigma(F)$') will result in different statistics. Checking the exact number of reflections will reveal this problem and others such as different implicit resolution cut-offs.
- The refinement is not progressing at all. Inaccurate standard deviations can confuse SHELXL as the standard deviations of the reflections are used in the weighting schemes.

10.2.3 Including data to atomic resolution

After the model has undergone a first round of corrections, the high-resolution data can be included. The `STIR` instruction is used to gradually introduce the new data

```
CGLS 20 -1      ! Run 20 cycles of conjugate gradient
                !  refinement
SHEL 10 0.1     ! Use data between 10 Å and full resolution
STIR 1.5 0.05 ! Begin refinement with data to 1.5 Å, then
                !  with every cycle include successive 0.1 Å
                !  shells of data
```

If the data extend to more than atomic resolution (e.g. $d_{\min} < 0.9$ Å), it may be more efficient to do some work on the model including only data to some intermediate resolution cut-off (say 1.0 Å) where many details can already been seen. The main reason for this approach is that computing times are significantly shorter for 1.0 Å data than for 0.8 Å data (the number of possible reflections approximately doubles when going from 1.0 Å data to data at 0.8 Å resolution).

10.2.4 Going anisotropic

As soon as data to at least 1.2 Å resolution have been included, anisotropic displacement parameters can be refined. Technically, this transition is triggered by adding the `ANIS` instruction to the ins-file.

When ADP's are introduced into the refinement, the value of R_{free} should be monitored closely. If the drop in R_{free} is less than 1–1.5%, it is better to revert to an isotropic refinement. As an intermediate solution, one can make only the heavy atoms in the structure anisotropic (e.g. sulfurs and/or metals) by applying the `ANIS` instruction only to the respective atoms (see example in Chapter 11 of the SHELXL manual).

10.2.5 Rebuilding the model at atomic resolution

Apart from the rebuilding of the protein and the solvent model in the same way as it is done for structures at less than atomic resolution, the major part of manual modeling at atomic resolution concerns the introduction of multiple discrete conformations. Accurate modeling of multiple conformations, in particular in active sites, often leads to much cleaner electron density and increased interpretability.

In this context, it should be stressed, that it is normally more efficient to wait with building multiple conformations until after anisotropic displacement parameters have been refined, as the inclusion of these parameters often leads to a dramatic improvement of map quality.

Manual rebuilding of models at atomic resolution

Concerning the rebuilding of incorrect parts of the model, there is still a lack of graphics software that is seamlessly integrated with SHELXL. Presently, the most convenient programs to use are Xfit (McRee, 1999) and Coot (Emsley and Cowtan, 2004). After modifications have been applied to the model, these programs write a pdb-file that then needs to be converted into an ins-file. This conversion can be done with SHELXPRO. However, to maintain full control over the parameterization of the model, one may need to be pragmatic and resort to using a text editor (the choice of the author of this chapter is Vi).

Before submitting an newly created ins-file for a complete run, it may be helpful to run one cycle of refinement against a fake reflection file containing only 100 reflections—such a run will not spend much time on analyzing the diffraction data but quickly generate the full geometrical analysis of the model which can be used to find out where mistakes have been made in the manual rebuilding.

Identification of regions to adjust
To identify regions of the structure that need rebuilding, the SHELXL lst-file contains a number of useful diagnostics:

- The List of 'Disagreeable restraints': After the last cycle of refinement this list contains all situations in which a restrained property of the model, be it a geometrical property such as a bond angle or an ADP-related property such as the directionality of ADPs of neighbouring atoms, is deviating by more than three times the r.m.s.d. specified for the restraint imposed on the property of interest. For problems related to ADPs a visual inspection of the corresponding ellipsoids (e.g. in Xfit) can be very revealing.
- The list of highest peaks in a $(1F_o - 1F_c)$ difference density map (search for 'Q1' in the lst-file): The `PLAN` instruction (e.g. `PLAN 200 -1 0.1`) will include the highest peaks into the pdb-file written by SHELXL to be conveniently located when the model is displayed. The positions of these peaks can also be identified in graphics programs by performing a peak search in the appropriate electron density map.
- Missing atoms: Atoms not included in the model can be identified from the entries in the list with the heading 'Following atoms could not be matched for particular residues for DFIX'.
- Unphysical Anisotropic Displacement parameters: Atoms for which the anisotropic displacement parameters have refined to unreasonable values (mathematically corresponding to hyperbolas instead of ellipsoids) are marked as non-positive definite (search for 'NON POSITIVE DEFINITE'). Atoms for which the ADPs describe very elongated ellipsoids are marked as 'maybe split'.

Sometimes, entries in the list of disagreeable restraints correspond to a real deviation between the model and the expected stereochemistry. For example, the ω-angle that describes the flatness of the peptide bond can deviate significantly from 180° depending on the actual electronic situation in a particular peptide bond (e.g. König *et al.*, 2003). However, in the majority of cases, disagreements between the model and the restraints imposed have their reason in the model being incorrect. In particular, multiple conformations that are modeled as one conformation leave a clear footprint in terms of violated `DELU` and `SIMU` restraints as the modeling of a site, which in reality is only 50% occupied, with an occupancy of 1.0 will be compensated by unrealistic shifts in the (anisotropic) B-value of the respective atom. This will result in unreasonably large differences between this B-value and the B-values of the neighbouring atoms.

In most cases, atoms that have been marked as non-positive definite should be removed from the model and possibly be reintroduced after some cycles of refinement if difference density persists.

Atoms with very elongated ellipsoids should be inspected. However, cases where it is appropriate to introduce two discrete conformations will usually be found when the residues mentioned in the list of disagreeable restraints are inspected.

Introduction of multiple conformations

When a region of a structure with signs of multiple discrete conformations has been identified, the most efficient strategy for inclusion of such multiple conformations into the model is a stepwise procedure, in which changes in the parameterization are followed by some cycles of refinement. The principal steps are (see Figure 10.2):

1. Constraining the occupancy of the existing conformation to a value of 0.65, whereby the rational behind using a value of 0.65 is that usually the major conformation will be detected first. It is also advisable to reset the B-values of the affected atoms to isotropic. After some cycles of refinement, the difference electron density of a potential second conformation will become significantly better defined.
2. Building of the second conformation into the improved electron density, assignment of PART numbers to the two conformations and application of the constraint that the sum of the occupancies of the two conformations must be one. After some cycles of refinement, the occupancy will refine to an optimum value with B-values that should not anymore give rise to disagreeable restraints. Once the parameters of both conformations have stabilized, the atoms can be made anisotropic again by using the `ANIS`-statement.
3. Definition of networks: If disordered groups of atoms are close to each other, there is usually a way to define mutually exclusive interpenetrating networks. These can be implemented by using the appropriate PART number and constraining the occupancies within the groups to a common value. Building the correct networks and constraining the occupancy values have only a small effect on the number of

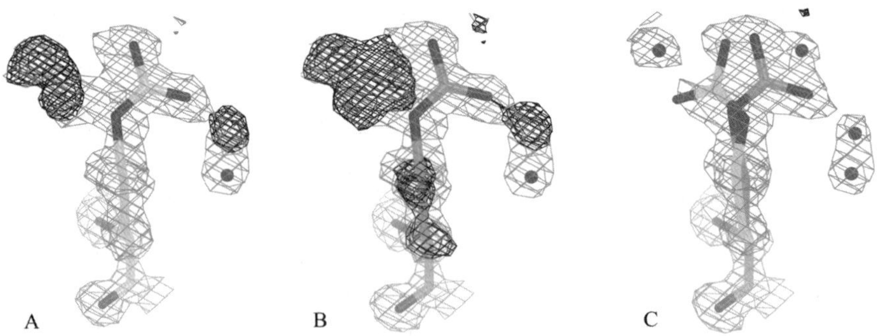

Fig. 10.2 Refinement of multiple conformations. In all panels, σA-weighted $2F_o - 1F_c$ difference electron density is shown at the 1σ-level (grey) and $F_o - F_c$ difference electron density is shown at the 2.5σ-level (black). The figure is based on a refinement of triclinic lysozyme at 1.1 Å resolution. The residue displayed is Arg45. A: One fully occupied conformer does not fully interpret the electron density. B: After reducing the occupancy of the sidechain atoms, the difference electron density becomes more interpretable. The sidechain atoms that should be kept at full occupancy (CB, CG, CD) exhibit positive $F_o - F_c$ difference electron density. C: After adding a second conformation and placing some corresponding water molecules in the remaining peaks of difference electron density, the model fully describes the electron density. The figure was created with PyMOL (DeLano, 2002).

parameters used in the refinement, but very frequently lead to a much more stable refinement and substantially improved electron density maps.

When occupancies refine to less than 20% it should be considered whether to go back to a description with only one conformer unless some atoms heavier than C, N, or O are involved (e.g. sulfurs) and/or large networks have been built.

Although SHELXL allows modeling three or more conformations (see Chapter 5 of this book), it is usually neither necessary nor possible to model more than two conformations in a stable manner. In this context, the description of lysine side chains with three or more conformations is a notoriously frustrating and—in the hands of the author—a never successful example. Atoms for which more than two conformers can occasionally be refined in a stable manner, are hydroxyl oxygens in serines and sulfur atoms in cystins and methionins.

The issue of what to do about atoms or sites for which no interpretable electron density can be found has been discussed at length (e.g. on the various macromolecular mailing lists and bulletin boards). There are still no generally accepted rules and all that can be done is to attempt to be consistent. The author of this chapter uses the following strategy:

1. If there is any density at all, at least one conformer is placed. The occupancy of its atoms are refined, or, if necessary, manually adjusted to fixed values, such that the refined values of the respective B-values are consistent with the B-values of neighbouring atoms.
2. If there is no density, the corresponding atoms are deleted from the pdb-file and a remark is attached to the final model to be submitted to the protein data bank. In principle, an occupancy of zero could be used to indicate atoms for whose location in the unit cell the data provide no evidence; however, many users of structural models are not aware of this mechanism and will be led to wrong conclusions.

A common observation for multiple conformations is that bonds and angles can be distorted for the atoms that are at the border between the ordered and the disordered part. A simple solution is to duplicate atoms from the ordered part such that strain can be released by small movement of the atoms in the different conformers. This strategy is, of course, at the expense of additional parameters. A typical example are CA-atoms of residues with disordered side chains; in many cases the positions of the CB atom will be slightly different between the different conformers, which, naturally, entails different positions of the CA and the other backbone atoms of the respective backbone atoms as well. Including more and more atoms into the disordered parts will successively resolve stereochemical problems but can lead to a 'zippering up' of large part of the structure which is not always desirable—some compromise has to be accepted in such cases.

Typical problems

• Especially after extensive rebuilding, SHELXL will sometimes stop with the message '*** REFINMENT UNSTABLE ***'. Often, this will be caused by newly constructed parts of the structures whose parameters undergo large shifts. These parts of the structure can be identified by checking for the maximum shifts in the lst-file (search for 'Max. shift') and then be removed from the model. In more subtle cases, the increased radius of convergence of running a round of refinement with the `STIR` instructions (poor man's simulated annealing) will remedy the situation.

• A common reason for unreasonable behaviour of a refinement job is a forgotten (closing) `PART 0` instruction, leading to the use of an incorrect occupancy for all atoms until the next `PART` instruction is encountered. This problem can be spotted by inspecting the pdb-file with a text editor.

• If during manual rebuilding the local geometry becames much distorted, resulting in an incorrect construction of SHELXL's internal connectivity table (the complete connectivity table is printed in the lst-file under the heading 'Covalent radii and connectivity table'), the `BIND` and the `FREE` instructions can be used to explicitly fix problems.

• When a second conformation is built into $1F_o - 1F_c$ difference electron density, the $2F_o - 1F_c$ difference density after some cycles of refinement does not reach the expected level. However, upon removal of the second conformation, the $1F_o - 1F_c$ density frequently reappears. To avoid endless cycling, a pragmatic criterion for accepting a second conformation is to not require the presence of $2F_o - 1F_c$ density but to require the absence of any signal at less than the -3σ-level in the $1F_o - 1F_c$ difference density map.

10.2.6 Inclusion of hydrogens—when and how

Hydrogen atoms are an important part of protein structures as they determine many of the non-bonded interactions, and in many cases they are the key players in catalytic events. Due to their small scattering power, hydrogen atoms are difficult to detect in electron density maps at less than atomic resolution. However, at atomic resolution many hydrogen atoms can be positively identified in difference electron density maps and then modeled with confidence.

From a technical point of view, the inclusion of hydrogens nearly doubles the number of atoms for which structure factor calculations have to be performed and thus leads to a significant cost in terms of computing time. Given that the inclusion of hydrogen usually only leads to a relatively small improvement in refinement statistics (typical drops in R_{work} and R_{free} are on the order of 0.5–1.0%) and clarity of the electron density maps, it is therefore advisable to only include hydrogen atoms at the later stages of refinement for the sake of efficiency.

For the refinement of protein structures, the SHELXL user does not need to understand how to use the full arsenal of parameterization for hydrogen atoms available (see the SHELXL Manual and Chapter 3 in this book). All `HFIX` instructions necessary to generate the hydrogens in a given protein structure can be conveniently generated by SHELXPRO and stored in the ins-file in the form of comments that allow activating the `HFIX` statements when required (by removing the preceding `REM` card). In practice, it has proven advisable to not activate the generation of all hydrogen atoms, but instead limit the hydrogen atoms to the ones whose position can be calculated based on the position of the neighbouring non-hydrogen atoms (riding-model, see Chapter 3). For hydrogen atoms that can be in variable positions, such as hydroxyl hydrogens, it is preferable to leave their generation out to avoid model bias and possible unfavourable geometric situations by incorrect placement (see example in Sheldrick and Schneider, 1997). In fact, even in very high-resolution structures with high quality data (e.g. Aldose Reductase, Howard *et al.* (2004)), many of the hydroxyl hydrogens never become fully visible in electron density maps, which may be due to their involvement in flexible hydrogen bonding networks such as flip-flop networks (Saenger *et al.*, 1982). Furthermore, all hydrogen atoms on the imidazole moiety of histidines can be left ungenerated; this way, the hydrogen atoms on the carbon atoms can be used to calibrate the electron density for the detection of the protonation of the nitrogen atoms. The `HFIX` instructions for generating the non-imidazole hydrogens on all histidine residues in a structure are:

```
HFIX_HIS 13 CA
HFIX_HIS 23 CB
HFIX_HIS 43 N
REM HFIX_HIS 43 ND1 CE1 CD2
```

Note that this set of statements will also generate the requested hydrogens for histidine residues with atoms in multiple conformations.

Concerning the actual positions of hydrogen atoms placed by SHELXL, some caution may be necessary when distances between these hydrogens and other atoms are measured as SHELXL places the hydrogen atoms into a position that is optimal for refinement against X-ray diffraction data. These positions do not correspond to the actual centre of the nucleus of the hydrogen atom (see also Chapter 3 of this book).

Typical problems

- A common source of confusion is the treatment of the N-terminal amino group of a protein. If the corresponding `HFIX` instructions had not been generated by SHELXPRO, a statement of the form `HFIX 33 N_1` needs to be added before any other `HFIX` statements in the ins-file.
- Missing atoms often create problems in terms of connectivity that needs to be fully defined for calculating the positions of riding hydrogen atoms. If, for example, the side chain of residue 26 has not been modeled beyond CB, SHELXL will complain, as it cannot calculate the positions of the CB-hydrogens without knowing the positions of the CG atom. Switching off the generation of the CB atoms by adding an `HFIX_26 0 CB` instruction will remedy the situation.

10.2.7 Solvent

Ordered solvent

As in standard refinements of macromolecules, the modeling of ordered solvent molecules is an iterative process in which it is important to adhere to some rules in order to arrive at a consistent description and to avoid endless cycling:

1. For fully occupied water molecules, an acceptable water molecule must have at least one hydrogen bond to another atom in the structure. Water molecules will be removed from the model when their *B*-values exceed a certain threshold

(e.g. 50 Å^2, the precise value depends on the temperature of data collection and data quality).

2. For partially occupied water molecules three categories can be defined:
   a. Partially occupied water molecules which make a hydrogen bond to at least one other site of the model. The site can be fully or partially occupied. In the latter case, the PART number of the new water should be chosen accordingly and its occupancy constrained appropriately.
   b. Groups of partially occupied waters: Often, elongated electron density for a solvent molecule can be modeled with two sites. Likewise, banana-shaped electron density that corresponds to two pairs of half-occupied water sites can be found quite frequently. Such situations can be modeled by filling the sites with atoms whose occupancy is fixed to 0.5.
   c. 'Lonesome' water molecules: It often helps to temporarily interpret peaks in the solvent region with partially occupied water molecules although no real meaning can be attached to these molecules. The placement of such molecules ('dummy atoms') may improve the phases and eventually may reveal partially occupied buffer molecules. If no physical or chemical meaning can be attached, these sites should be removed when the model is finalized.

Bulk solvent

The bulk solvent model implemented in SHELXL is rather simple (see the SHELXL Manual). As a consequence, the agreement between observed and calculated diffraction intensities may not be as good at low resolution as at high resolution.

However, in many cases a disagreement between data and model will also have causes in experimental data. Inaccurate measurements of low resolution data can for example be due to overloaded reflections, reflections measured behind the beam-stop, synchronization problems due to excessively fast rotations of the crystal (stepping motors missing steps) combined with short exposure times (inaccurate opening and closing times of shutters). Some of these problems can be remedied by careful reprocessing of the data. If this is not possible, a low-resolution cut-off can be applied using the `SHEL` instruction. When the refinement is restarted with corrected diffraction data, the parameters of the solvent model should be reset to their default values by using a `SWAT` instruction without any parameters. Such a reset of the bulk solvent parameters can also become necessary when the `SWAT` parameters have been drifting out of their physically reasonable range (the first parameter, g, should be between 0.7 and 1.0; the second parameter, U, should be between 2 and 5, for more details see the SHELXL manual).

10.2.8 Finalizing the model

As with most crystallographic refinements, it is difficult to decide when a refinement at atomic resolution can be considered as finished. Many refinements of biological macromolecules are stopped when the biological question can be answered and/or the R_{work} is less than 20%. However, it should be kept in mind that in order to have an accurate description, for example of the active site, the model must be refined to completion also in other parts of the structure. Otherwise bias effects originating from inaccurate modeling of 'uninteresting' regions of a structure can lead to artifacts in the regions of interests.

Unit cell parameters

The unit cell parameters have to be known accurately to allow the measurements of distances on an absolute scale. Inaccuracies typically originate from errors in the values for the wavelength of the radiation used for the experiment (for synchrotron data) or from incorrect values for the distance between the crystal and the detector. Programs such as WHAT_CHECK (Hooft *et al.*, 1996) are available to detect such problems and to suggest corrected values of the unit cell parameters. When the unit cell parameters have been corrected, the structure must be adjusted to the new unit cell by some cycles of refinement.

Restraints

In principle, atomic resolution data contain sufficient information to support a refinement of the ordered parts of a macromolecular structure without restraints. However, in most real cases, it is not advisable to switch off the restraints or to alter their weights (using the `DEFS` instruction) as the restraints are absolutely necessary to keep the less ordered parts of the structure in check while in the more ordered parts of the structure, if appropriate, the data will move the model to the correct place even if restraints have been imposed.

However, some exceptions can be made. An instructive example in this context is the ω-angles describing the peptide bond. Even if restraints are applied, many of the ω-angles will assume values different from 0 or 180 degrees (showing that the data are moving the model away from the target value of the restraint). Deviations of up to 30°, which are in perfect agreement with electron density have been observed (e.g. see König *et al.*, 2003) and are in fact physically reasonable (MacArthur and Thornton, 1996). To obtain accurate values for the ω-angles, it may be required to remove the restraints describing the flatness of the well-ordered peptide bonds while explicitly keeping the restraints for the less-ordered peptide bonds. Another situation of interest is the refinement of carboxylate groups, where a difference between the lengths of the two C—O bonds can be used to infer protonation. Here, it may be better to switch off the standard restraint (that imposes equal values for both bond lengths). Again, this approach will only deliver useful results if the carboxylate group in question is in a well-ordered part of the structure.

Criteria for a final model

In addition to the criteria for a final model applied to medium and low resolution structures (e.g. agreement of the backbone dihedral angles with the Ramachandran plot, reasonable B-values, etc.), for a model refined at atomic resolution, a number of criteria can be given (see also Chapter 11 in this book):

1. There should not be any entries left in the 'list of disagreeable restraints' that could be resolved by adjusting the parameterization of the model. Acceptable exceptions

are for example situations where, in principle, more than two conformations should be included, but for technical reasons only two have been modeled.
2. The list of atoms without restraints should only contain atoms for which the restraints have been intentionally removed.
3. The description of disorder should be as consistent as possible, that means networks should be built wherever possible.
4. No atom should be marked as non-positive definite.
5. The r.m.s.d. of the $1F_o - 1F_c$ difference electron density map should be around 0.07 to 0.1 electrons/A^3.
6. No significant peaks (positive or negative) should remain in difference electron density maps, whereby a pragmatic definition of a significance threshold will lie somewhere between 4.5 and 5 σ (where σ is the r.m.s.d. of the $1F_o - 1F_c$ difference electron density map). Remaining peaks are acceptable if a plausible interpretation such as 'probably a third conformation of a Ser, but not modeled' can be given.
7. Hydrogen atoms must be placed in a complete and consistent manner.

Technical aspects

- To achieve a consistent placement of hydrogen atoms, a simple approach is to first delete all hydrogen atoms from a model and then regenerate a new set of hydrogen atoms using `HFIX` cards as generated by SHELXPRO (see Example 10.3.1).
- When the final parameterization of the model has been reached, the reflections of the test set have to be included. This can be affected by removing the '-1' argument from the `CGLS` instruction. Should problems appear in the resulting model, sites or parameters can be removed, but, by no means should parameters be added at this stage.

10.2.9 Estimation of coordinate uncertainties

Due to the large number of observables in an atomic resolution refinement, the inversion of the normal matrix of the refinement can be used to estimate standard uncertainties for the refined parameters (Cruickshank, 1970; Press *et al.*, 2002). The corresponding calculations are enormous and until recently could only performed on large mainframe-type computers. However, with the rapid progress in computing technology, the matrix inversion can now be done on standard crystallographic workstations in a couple of hours.[3]

For many structures of macromolecules at atomic resolution presented in the literature, all restraints were removed from the target function before the inversion of the normal matrix. However, recently it has been shown that in regions of the model where the restraints are important, for example when two sites belonging to two different conformations share the same electron density (König *et al.*, 2003), the estimated standard uncertainties can be overestimated. Nevertheless, the estimated standard uncertainties for well-separated sites will be reliable.

[3] Inversion of a matrix for 7433 parameters against 41006 reflections for Tendamistat (König *et al.*, 2003) used 138 MB of memory and took 33 min. of CPU time on a Intel P4 processor running at 2.2 GHz under Linux.

Technical aspects

- To prepare an ins-file for a matrix inversion job, all restraints should be removed and the shift multiplication parameters be set to zero (`DAMP 0 0`). The successful removal of all restraints can be checked by looking at the number of restraints counted in a test job. In polar space groups, one restraint that fixes the origin will need to remain.
- If the numerical problem is too large to be solved by a particular version of SHELXL, the program will complain with error messages such as '***** ARRAY B TOO SMALL FOR THIS PROBLEM ***'. One can then use a precompiled larger version of the program, called SHELXH, or recompile the program with increased array dimensions (http://shelx.uni-ac.gwdg.de/SHELX/#Compilation). Another option is to reduce the size of the problem by limiting the inversion to parts of the normal matrix that correspond to certain parameters, such as coordinates, using the `BLOC`-instruction (see the SHELXL manual for details).
- If one is interested in the standard uncertainties of quantities that are derived from refined parameters such as bond lengths and angles, `RTAB` instructions (see Chapter 2) can be added to trigger the calculation of their values and their estimated standard uncertainties as derived by error propagation (which is based on the full variance-covariance matrix of the problem). For details see Example 10.3.2.

Typical problems

- Inversion of normal matrix can become numerically instable resulting in 'hanging' jobs. Often such instabilities are caused by atoms with zero occupancy, which, in fact, can be removed from the model (with the complication that hydrogens attached to those particular atoms also need to be reorganized). In other cases it may be necessary to go back to the last CGLS-based refinement job to check for any parameters that are still shifting or oscillating.

10.2.10 Analysis and presentation of the structure

A structure at atomic resolution gives a much more detailed picture of a macromolecule than for example a structure at medium resolution. On one hand the higher resolution leads to smaller details becoming visible in electron density maps. On the other hand the much more elaborate parameterization of the model allows answering qualitatively new questions, like whether the anisotropic ellipsoids of two atoms in an active site are pointing towards each other.

Before going into interpretation of the structural data, the overall quality of the model needs to be assessed. In addition to the standard statistics provided for models at lower resolution (R_{work}, R_{free}, R_{all}, agreement of the model with stereochemical restraints, mean B-values, etc.), quantities that characterize the atomic resolution model such as agreement of the anisotropic displacement parameters with the restraints imposed, must be quoted. Some interesting statistics on ADPs can be obtained with Ethan Merritt's PARVATI-server (Merritt, 1999).

Due to the strength of the diffraction data, structures at atomic resolution may have more incidences of 'abnormal' values as normally expected by the validation programs for a well-refined structure at medium resolution. Typical examples are extreme values for ω-angles; at medium resolution, an omega angle of 155° would hardly be believable while at atomic resolution, the electron density proves the point (König *et al.*, 2003). Such deviations should be discussed in a publication.

A very instructive way of displaying anisotropic displacement parameters is to plot the corresponding vibrational ellipsoids at some given level of probability ('ORTEP-plots' in small molecule crystallography). Such plots allow an intuitive assessment of the correctness and the meaning of the ADPs. For macromolecules, vibrational ellipsoids can be displayed using Xfit (McRee, 1999) and BOBSCRIPT (Esnouf, 1999). By application of Rosenfield's rigid-body criterion (Rosenfield *et al.*, 1978), ADPs can also be used to characterize the flexibility of a molecule as described in Schneider (1996a).

To display disordered regions of a molecule with display programs that do not properly handle different PART numbers, it may be necessary to split the pdb-file into two copies, one containing `PART 0` and `PART 1`, the other one `PART 0` and `PART 2` and then work with the two copies.

Given that the current technologies used for the refinement of protein structures at atomic resolution can be seen as either standard protein techniques being expanded to high resolution or as standard small molecule techniques expanded to somewhat lower resolution, there is still much room for qualitatively new developments (one example being the use of normal modes to model the anisotropic displacements of atoms in a crystal (Kidera *et al.*, 1992)). This is one reason why it is of particular importance to deposit the experimental data together with the refined model, so that future generations are in a position to extract more information from them.

10.3 Examples

10.3.1 Course of a typical refinement of a protein

A typical example of a refinement of a small protein at atomic resolution is the refinement of Tendamistat, a protein consisting of 74 residues at 0.93 Å (König *et al.*, 2003). All intermediate files for the refinement of Tendamistat are included on the CD-ROM accompanying the book. The course of the refinement is summarized in Table 10.2.

The data were processed with DENZO and SCALEPACK (Otwinowski and Minor, 1997), divided into a work and a test set, and prepared for use in SHELXL (Sheldrick, 1997b) with XPREP (Sheldrick, 2001). A starting model was obtained by providing 4 sulfur sites taken from a structure of Tendamistat previously refined to 2.0 Å resolution (Pflugrath *et al.*, 1986; pdb-code 1HOE) to SHELXD and expanding the sulfur sites to a total of 634 atoms using dual-space recycling methods. This model was then submitted to arpWarp (Perrakis *et al.*, 1999) to obtain an initial model of the protein by automatic interpretation of the electron density based on diffraction data truncated to 1.5 Å resolution. After 50 cycles with default parameters, arpWarp had automatically built 73 of the 74 residues.

Table 10.2 Refinement of Tendamistat with SHELXL

Name	N_{obs}	N_{par}	R_{work} [%]	R_{free} [%]	Action taken
tenda1	9,985	2539	18.44	20.36	Initial model
tenda2	38,997	2579	19.28	20.19	MOD
tenda3	38,997	5799	14.03	15.69	Anisotropic Displacement Parameters
tenda4	38,997	5974	13.54	15.33	FOCC, WAT
tenda5	38,997	5974	13.16	14.96	DOUBLE, FOCC, WAT
tenda6	38,997	6002	12.85	14.50	DOUBLE, FOCC, WAT
tenda7	38,997	6191	12.61	14.35	DOUBLE, FOCC, WAT
tenda8	38,997	6191	11.80	13.27	Activate hydrogens
tenda9	38,997	6464	11.67	13.26	DOUBLE, FOCC, WAT, glycerol
tenda10	38,997	6519	11.26	12.55	WAT
tenda11	38,997	6574	10.90	12.24	WAT
tenda12	38,997	6842	10.52	12.13	WAT
tenda12_cgls	41,050	6842	10.50	n/a	Refine against all data

Name corresponds to the name of the job on the CD-ROM. N_{obs} is the number of reflections against which the structure has been refined. N_{par} is the number of parameters refined in the respective model. R_{work} and R_{free} are the R-values for $F > 4\sigma(F)$ calculated against work and test-set, respectively. Explanations for 'Action taken' are given in the text.

The resulting model was converted from pdb to ins-format with SHELXPRO and refined with isotropic B-values against data to 1.5 Å with SHELXL (tenda1). After some minor changes to the model, all data were gradually included into the refinement using the `STIR`-statement (tenda2). The use of anisotropic instead of isotropic displacement parameters more than double the refined parameters and lead to a drop in both R_{work} and R_{free} of 5.2 and 4.5%, respectively (tenda3). For the following four rounds of refinement (until tenda7), multiple conformations were modeled by fixing the occupancy of the first conformation to 0.65 (FOCC) followed by adding the second conformation (DOUBLE), and by including water molecules (WAT). Then hydrogen atoms were included in the model (tenda8) resulting in a drop in R_{work} and R_{free} of 0.8% and 0.9%, respectively. This was followed by three more rounds of modifications and refinement in which networks of multiple conformations were built and a glycerol molecule in the solvent was added. The refinement converged at R_{work} of 10.5% and R_{free} of 12.1%. To prepare the final model, all hydrogens were removed (tenda12_noh on the CD-ROM) and freshly placed. The resulting model was then subjected to 20 final cycles of conjugate gradient refinement (tenda12_cgls) against all data. Finally, estimated standard uncertainties were determined by inversion of the full matrix (tenda12_ls on the CD).

10.3.2 Determination of standard uncertainties for protein-ligand contacts

Use of `RTAB` statements will trigger the evaluation of estimated standard uncertainties for structural properties of interest. Following is an excerpt from the `RTAB` statement used to generate a table of protein ligand distances together with their respective standard uncertainties in the analysis of the ultra-high resolution structure of Aldose Reductase (Howard *et al.*, 2004):

```
RTAB INHI C2_320   NE1_20
RTAB INHI BR8_320  OG1_113
RTAB INHI F9_320   O_47
RTAB INHI F14_320  CH2_111
RTAB INHI F14_320  N_299
```

In the lst-file, the results of the calculation are presented as follows:

```
Distance INHI
      3.1378 (0.0040)  Lig_320  C2_320 - NE1_20
      2.9727 (0.0032)  Lig_320  Br8_320 - OG1_113
      3.0096 (0.0037)  Lig_320  F9_320 - O_47
      3.2245 (0.0043)  Lig_320  F14_320 - CH2_111
      3.2627 (0.0047)  Lig_320  F14_320 - N_299
```

The first column contains the value of the property measured (in this case the interatomic distance); the corresponding standard deviations are given in parentheses in the second column.

11

Protein structure (cross) validation

Michael R. Sawaya

Cross validation is a key concept in attaining accuracy in crystallographic refinement. Refinement of protein structures is especially subject to model bias; an unchecked imagination during the model building process or an inattentiveness to detail can reinforce features in the electron density map that are incorrect. The problem of model bias stems from the fact that protein structure refinement is almost always an underdetermined problem; our initial estimates of the phases are poor and discarded quickly, and we do not have a sufficient number of structure factors to justify the number of parameters refined. Prior to 1992, protein crystallographers relied on two simple criteria, obtaining an R-factor below 20% and a model geometry deviating less than 0.02 Å and 3° in bond lengths and angles, respectively. Though certainly necessary, these criteria proved to be disturbingly ineffective to guarantee model accuracy. Indeed, several instances of gross error in tracing the chain were reported in the literature, prompting the introduction and development of *cross* validation methods—methods that gauge a structure's likelihood to be correct based on criterion *not* used in the refinement process.

One of the most useful and widely accepted cross validation methods is the use of R_{free} (Brünger, 1992). A subset of reflections are withheld from the minimization process and only checked periodically for their agreement to F_c. If R_{free} drops during the refinement, the crystallographer obtains an unbiased indication that the refinement has gone well and no gross errors have been introduced in the model.

Numerous other cross validation methods have been developed to measure the geometric quality of the model coordinates, whereas R_{free} measures the agreement between F_c and F_o. Authors of structure validation programs have devised clever methods to parameterize the essence of an ideal protein model. In all the cases considered here, correlations are calculated between a given structure model and geometric parameters extracted from a library of very well-refined, high resolution models. The key concept is that the parameters used for the comparison were not used by the refinement program's minimization algorithm.

In the sections that follow, the algorithms of PROCHECK, WHATCHECK, Verify3D, ERRAT, and PROVE will be described individually, and program outputs illustrated and interpreted for the particular case, PDB ID code 1ja3 (Dimasi *et al.*, 2002). After surveying all the structures deposited in the PDB, 1ja3 was flagged

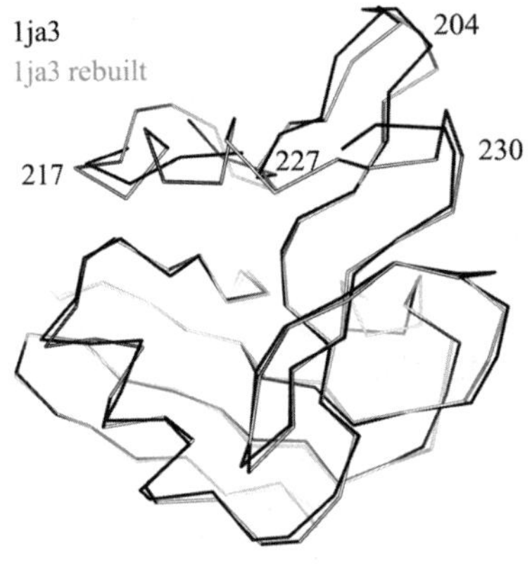

Fig. 11.1 Backbone representation of the model of 1ja3 as deposited with the PDB (black) and after revision based on the output of the various cross validation programs (grey). The region in the upper third of the figure (residues 204–230) required the most rebuilding.

by ERRAT as having particularly poor statistics. The structure's shortcomings are echoed loudly by the other validation programs as well. Troubles were found to extend over the top face of the protein (Figure 11.1, residues 204–230), in particular, there is a stretch of 15 residues (residues 217–230) modeled incorrectly as a broken β-strand. Using the structure validation tools and model rebuilding strategies described here, it was found that this latter sub-region could be more accurately modeled as a helix. Improvements in the model (including this stretch and other isolated areas) led to a drop in the $R_{\rm work}$ from 27.8% to 24.7% and the $R_{\rm free}$ from 28.3% to 27.4%. Readers may compare program output for both the deposited model and the model after rebuilding. This example is particularly informative because the medium resolution data (3.0 Å) typifies most ordinary protein crystals. It also illustrates how far a structure model can deviate from reality if cross validation is not employed. The .pdb files of all models as well as all output files of the validation programs are to be found on the CD-ROM that accompanies this book.[1]

The structure validation programs described here can be accessed individually (web addresses given in the References section at the back of this book), or conveniently all together using the Structure Analysis and Validation Server (SAVS) at http://nihserver.mbi.ucla.edu/SAVS.

11.1 PROCHECK

The PROCHECK structure validation program (Laskowski *et al.*, 1993) deploys a multitude of quality checks; using both cross-validation and standard checks for geometric deviations. The Ramachandran plot it produces is its most attractive and useful feature (Ramakrishnan and Ramachandran, 1965). The plot consists of a two dimensional plot of phi/psi values for each amino acid residue. Steric overlap between side chain and main chain atoms limits the energetically allowed values

[1] The file 1ja3_start.pdb contains the model as deposited with the PDB, 1ja3_final.pdb corresponds to the model after all modifications and the other .pdb files document the way from the one to the other model. The validation program output files can be found in the folder valid-output.

of (phi/psi). Hence, residues falling outside the allowed regions (marked by yellow and red colours on the plot) should be examined more closely. If a residue falls in a disallowed region, its residue number will be labeled. Evaluation of the plot can be considered a method of cross-validation because phi and psi angles are generally not optimized by automated refinement programs. A well-refined structure typically has 80–90% of residues in the 'most favoured' region and the remaining residues in the 'additionally allowed' region.

The Ramachandran plot is helpful in identifying localized errors involving a single amino acid (a single Ramachandran outlier) or more global problems such as tracing the chain in the wrong direction (multiple outliers consecutive in sequence). In most cases outliers on the Ramachandran plot correspond to isolated amino acids requiring a peptide flip (that means rotating the peptide plane 180° so the carbonyl oxygen points in the opposite direction), which means that the main chain and side chains are positioned correctly (atoms CA, CB, etc.), but the peptide plane (atoms N, C, O) is oriented incorrectly. At map resolutions worse than 2.5 Å, this error becomes increasingly common due to the inability to see the carbonyl bump (bulge due to the protuberance of the carbonyl oxygen) in the map (as in Figure 11.2). If there are numerous consecutive outliers in a region of the map with high *B*-factors (such as a solvent exposed loop), one could consider using the library of loops available in the 'O' graphics package (lego_loop) to help chose how the loop should be modeled (Jones *et al.*, 1991). The library is taken from well-refined, high resolution structures, and thus should have good geometry.

Disallowed phi/psi angles are rare in proteins, but there are exceptions. The ultimate guiding factor should be the electron density map. If the map clearly indicates that the carbonyl oxygen is correctly modeled and no other interpretation seems reasonable, then the phi/psi angles should not be changed. Sometimes disallowed phi/psi angles are key features of the protein structure providing some structure or functionality that would otherwise be impossible to achieve. In general, if the disallowed phi/psi angle is a true feature of the protein structure, then there will be a network of hydrogen bonds or other favourable interactions with the peptide backbone to stabilize its strained conformation. Also, there should be no strained bond lengths or bond angles reported for residues involved. To check for these deviations the last page of PROCHECK output should be consulted, as discussed below.

A glance at the Ramachandran plot for 1ja3 (Figure 11.2), suggests that this model contains serious errors affecting the path of the main-chain. Only 65% of the residues lie in the 'most favourable' region of the plot, 30% percent of the residues lie in the 'additionally allowed' regions, and the remaining 5% of the residues lie in the 'generously allowed' region of the plot (labeled with residue number). One of the outliers, Phe260, is highlighted in Figure 11.2. Because the diffraction data extend to only 3.0 Å resolution, there is no carbonyl bump in the electron density to guide the positioning of the backbone carbonyl oxygen. After careful examination of the $2F_{\mathrm{O}}-1F_{\mathrm{C}}$ electron density map, it was observed that both phi and psi should be flipped and an additional residue introduced to fill the density. The new conformation for Phe260 placed it in one of the 'most favourable' regions of the Ramachandran plot.

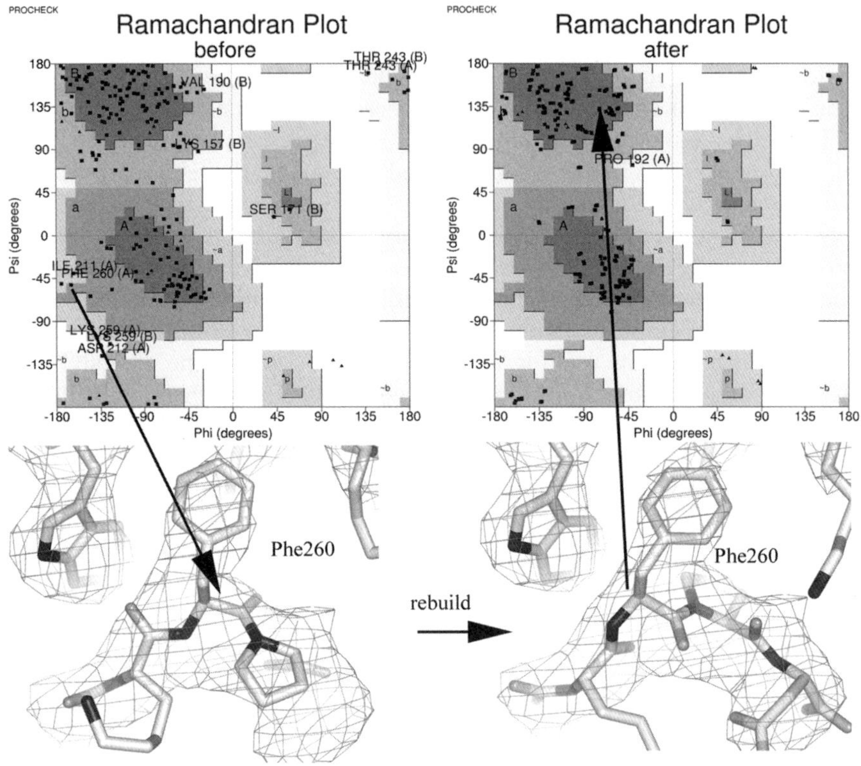

Fig. 11.2 Ramachandran Plot generated by PROCHECK for the model of 1ja3 as deposited with the PDB (left) and after revision (right). Residue 260 moves from the 'generously allowed' region of the Ramachandran Plot to the 'most favourable' region after remodeling.

The remainder of the PROCHECK output evaluates bond length, bond angle, planarity of aromatic and amide groups. These deviations should normally be quite small since automated refinement programs restrain these parameters to acceptable values. If the output reports numerous outliers, the weighting scheme in the automated refinement program can be changed to enforce more ideal geometry. If the output reports a few outliers, these should be checked individually. They are especially significant if the outliers are consecutive residues in the chain. Usually, an alternate interpretation of the density is possible which will alleviate the strain. Oftentimes the error is accompanied by peaks in the $F_o - F_c$ map. In the 1ja3 example, 10 main-chain bond angles were flagged as exceeding 10° from ideal. Five of these violations are clustered in a localized area, involving residues 216, 218, 219, 221, and 222. These residues also correspond to the most offensive violations of the Verify3D and ERRAT plots (Figures 11.4 and 11.5) discussed below. Because so many residues were involved, it was decided to remove the residues from the model and calculate an omit map (Figure 11.3). After careful examination, it was decided that the region should be changed from a β-strand to an α-helix.

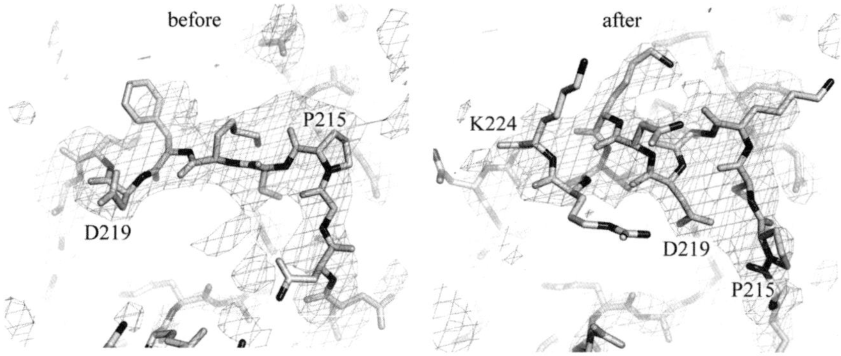

Fig. 11.3 An omit map contoured at 1.2 σ for the model of 1ja3 as deposited with the PDB. The left panel illustrates the starting model, containing a broken β-strand. The right panel is the revised model.

PROCHECK can be downloaded from the following web site: http://www.biochem.ucl.ac.uk/~roman/procheck/procheck.html. To run PROCHECK, one simply types 'procheck mycoordinates.pdb resolution_limit', where mycoordinates.pdb should contain coordinates of your refined protein structure in standard Protein Data Bank (PDB) format, and resolution_limit should correspond to the high resolution limit of the data (in Å) used in the refinement. Alternatively, one can submit coordinates to either of two websites: the SAVS server or the PDB validation server (www.deposit.pdb.org/validate/). For additional information, consult the PROCHECK manual www.biochem.ucl.ac.uk/~roman/procheck/manual/.

11.2 WHAT_CHECK

WHAT_CHECK structure validation program (Hooft *et al.*, 1996) checks approximately 40 different features of a protein's geometry. These checks are aimed at catching all levels of errors, from the careless omission of a protein molecule in the asymmetric unit to the subtle nuances of side chain atom nomenclature. It is easy to get lost in the volumes of information output by WHAT_CHECK. Special attention should be devoted to the following three key checks.

11.2.1 List of close non-bonded contacts

Atoms should not approach each other closer than the sum of their van der Waals contacts. Often the worst offenders are atoms in disordered regions of the map. Because the map is not well-defined in these regions, it is usually possible to choose an alternate rotamer that avoids the steric clash yet still fits within the density envelop. If the offending atoms are main chain atoms, one could consider using the library of loops available in the 'O' graphics package (Jones *et al.*, 1991). Keep in mind that numerous close contacts between a water molecule and neighbouring carbonyl oxygen atoms might signify that the modeled water molecule is really a metal ion.

Examination of the output of 1ja3 reveals numerous close contacts, the worst of which is over 1 Å closer than the sum of the van der Waals radii:

```
      Residue i                          Residue j         Distance  Distance
type number chain atom      type number chain atom violation observed
SER  ( 216 ) A     CA   --  LYS  ( 217 ) A     CE   1.060      2.140 INTRA BF
SER  ( 216 ) A     C    --  LYS  ( 217 ) A     CE   0.911      2.289 INTRA BF
SER  ( 216 ) A     CA   --  LYS  ( 217 ) A     NZ   0.885      2.215 INTRA BF
LYS  ( 221 ) A     NZ   --  SER  ( 229 ) A     N    0.807      2.193 INTRA
VAL  ( 140 ) A     CG1  --  MET  ( 155 ) A     CG   0.637      2.563 INTRA
SER  ( 216 ) A     C    --  LYS  ( 217 ) A     CD   0.631      2.569 INTRA BF
PRO  ( 215 ) A     O    --  LYS  ( 217 ) A     CE   0.594      2.206 INTRA BF
ARG  ( 230 ) A     CG   --  GLY  ( 231 ) A     N    0.502      2.598 INTRA BF
CYS  ( 163 ) A     O    --  LYS  ( 164 ) A     C    0.485      2.315 INTRA BF
GLY  ( 214 ) A     O    --  SER  ( 216 ) A     N    0.460      2.240 INTRA BF
PRO  ( 215 ) A     C    --  LYS  ( 217 ) A     CE   0.419      2.781 INTRA BF
LYS  ( 206 ) A     N    --  GLU  ( 207 ) A     N    0.367      2.233 INTRA BF
PHE  ( 260 ) A     N    --  PRO  ( 261 ) A     CD   0.358      2.642 INTRA BF
SER  ( 216 ) A     N    --  LYS  ( 217 ) A     CE   0.325      2.775 INTRA BF
LYS  ( 203 ) A     O    --  LYS  ( 205 ) A     N    0.320      2.380 INTRA BF
THR  ( 159 ) A     O    --  GLY  ( 162 ) A     N    0.320      2.230 INTRA BF
LYS  ( 203 ) A     C    --  LYS  ( 205 ) A     N    0.275      2.625 INTRA BF
TRP  ( 160 ) A     C    --  GLY  ( 162 ) A     N    0.261      2.639 INTRA BF
                                 And so on for a total of 114 lines
```

Notice that 9 of the top 18 offenders are again in a localized cluster (residues 216–221) corresponding to the incorrectly modeled β-strand mentioned in the previous two sections.

11.2.2 Unsatisfied hydrogen bond donors/acceptors

At the resolution limits of most protein structures it is impossible to distinguish between nitrogen and oxygen atoms simply by the height of the electron density peaks. Instead, it is necessary to examine the environment of the atoms. It is unlikely to find two hydrogen bond donors within hydrogen bonding distance (2.3–3.2 Å), nor is it likely to find two hydrogen bond acceptors within such a short distance. In such cases, one should check whether it is possible to flip a nearby amide group to exchange the positions of nitrogen and oxygen atoms (for example flipping the amide of Asn or Gln, or the imidazole ring of His). Specifically, see output message. 'Error: HIS, ASN, GLN side chain flips' for a list of side chains that might require flipping to optimize hydrogen bond networks.

In the 1ja3 example, the results appear much more serious than a simple side-chain flip. Most of the unsatisfied hydrogen bond donors/acceptors are main-chain backbone amides:

```
ASN   ( 156 ) A    N
THR   ( 159 ) A    N
LYS   ( 176 ) A    N
ILE   ( 177 ) A    N
ILE   ( 191 ) A    N
```

```
ILE   ( 197 ) A   N
LEU   ( 199 ) A   N
LYS   ( 204 ) A   N
GLU   ( 207 ) A   N
TRP   ( 210 ) A   N
ILE   ( 211 ) A   N
ASP   ( 212 ) A   N
GLY   ( 214 ) A   N
SER   ( 216 ) A   N
SER   ( 216 ) A   OG
ILE   ( 247 ) A   N
ILE   ( 252 ) A   N
```

Again, many of these residues are localized to the most troubled region of the structure (residues 204–230). After rebuilding, many of these residues were eliminated from the list, though not all.

11.2.3 List of isolated water molecules

Water molecules are never ordered unless they are hydrogen bonded to another atom. In such a case, it would be best to remove the orphaned water molecule from the model and run another round of automated refinement. Examine the $F_o - F_c$ map in this region after the refinement. Chances are that the density will have disappeared, confirming that the water was modeled incorrectly. If, however, the positive density is found, then one should broaden one's view and look for the possibility that the density arose from an alternate side chain conformation or a larger ligand. If the resolution of the data is worse than 2.7 Å, there are usually little or no water molecules to be found, as is the case in the 1ja3 example.

11.3 Verify3D

Verify3D (Lüthy *et al.*, 1992) is an effective tool for detecting global errors in a protein structure, such as whether the chain has been traced in the wrong direction or if there is a sequence registration error. The program examines how compatible the three-dimensional structure is with the primary structure. Each of the twenty amino acids is given three parameters describing its preference for (1) secondary structure, (2) degree of buried surface area, and (3) fraction of side-chain area that is covered by polar atoms. These three parameters are evaluated for each residue in the structure and a correlation is calculated between this set of observed parameters and the 'ideal' parameters of the amino acid type to which it has been assigned. For example, if an amino acid residue in the structure has been assigned as a leucine residue (an amino acid characterized by its preference to be buried and shielded from polar atoms) but the atoms in this residue are largely solvent exposed, the residue would receive a poor correlation score. Scores are averaged over a 21 residue window and plotted over the whole residue range. This algorithm can be considered a cross-validation

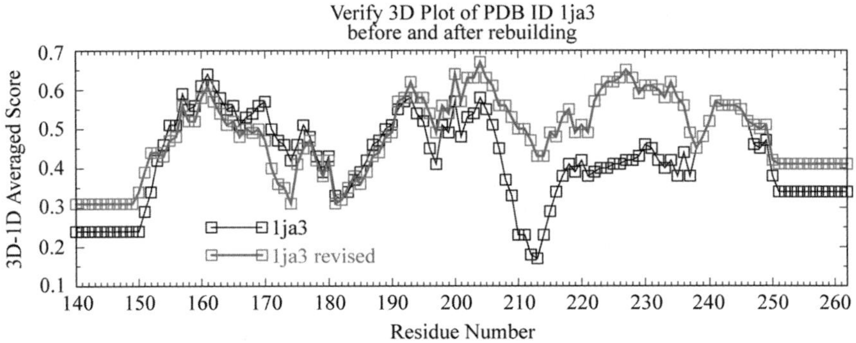

Fig. 11.4 Verify 3D Plot for the model of 1ja3 as deposited with the PDB (black) and after revision (grey). Notice the improvement in residues 205–235.

method since automated refinement programs do not evaluate these parameters. If a section of the plot dips below a correlation of 0.2, one should re-evaluate the sequence assignment.

In the 1ja3 example, it is again noted that the residues displaying the worst 3D-1D averaged scores are localized (residues 208–219) and correspond to the incorrectly modeled β-strand (Figure 11.4). After rebuilding this region into a helix, the 3D environments of these residues changed dramatically. The Verify3D plot shows significant improvement in the localized region and overall.

11.4 Errat

The Errat program (Colovos and Yeates, 1993) is exquisitely sensitive in detecting unusual atomic environments in protein molecules, employing an algorithm unlike any other validation program. The algorithm operates on the observation that the distribution of non-bonded, pair wise interactions between carbon, nitrogen and oxygen atoms is not simply random, but influenced by energetic and geometric effects imposed by protein molecules. A library of reliable, high resolution structures was evaluated for the frequency of each pair-wise (atom–atom) interaction type (C—C, C—N, C—O, N—N, N—O, and O—O) within cutoff distances (3.0–3.75 Å). The distribution of frequencies (that means the fraction of interactions contributed by each of the six types of atom–atom pairs) was found to differ significantly from what would be expected by a random collection of atoms. This empirically derived atomic distribution was used as the basis for statistically discriminating between correctly and incorrectly modeled regions of a query protein structure. Incorrectly modeled regions contain atomic distributions that approximate random values, whereas correctly modeled regions approximate library values. Scores are calculated over contiguous nine-residue windows indicating the confidence level that the model is in error in this region. The overall Errat score given for a structure signifies the percentage of residues falling below the 95% confidence limit. Most quality structures are 80–100% below the 95% confidence limit. If an individual

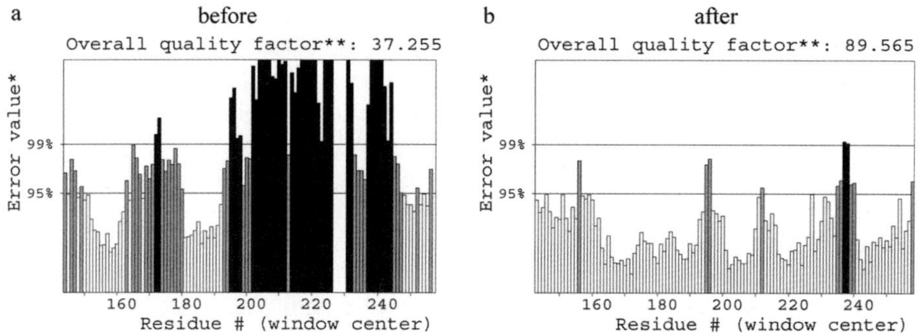

Fig. 11.5 Errat Plot for the model of 1ja3 as deposited with the PDB (**a**) and after revision (**b**). Interpretation of residues 210 to 220 as α-helix rather than as β-sheet dramatically improves the model and the Errat score.

window scores higher than 95% confidence level, a black bar appears in the graph over the window's central residue. Such an area should be examined more closely in the electron density map. The program may be considered a cross validation method because automated refinement programs do not evaluate atomic environments other than maintaining a proper van der Waals distance.

Errat has uncovered multiple instances of gross errors in structures deposited with the Protein Data Bank (PDB) (Colovos and Yeates, 1993). 1ja3 is one of the worst, having an Errat score of only 37.8. In other words, 63.2% of the structure was above the 95% confidence limit of containing an error (Figure 11.5A). Again, the most strongly offending region (residues 210–220) corresponds to the incorrectly modeled β-strand. After consulting an omit map (Figure 11.5B) the region was modeled as an α-helix (Figure 11.3). The relatively poorly featured electron density map is a result of the high overall Wilson B factor (68 Å^2), and was probably a major factor contributing to the map's original misinterpretation. The final Errat score improved from 37.8 to 88.8.

In other cases, the reason for the poor score is not always evident upon looking at the map. In these cases the problem usually lies in an incorrect assignment of the atom type. For example, a histidine residue might be flagged as incorrect by Errat because the CD atom of the imidazole ring has an unusual environment. The fit to the electron density may look fine, but there is something amiss about the chemistry of its environment. Perhaps a nearby oxygen atom is within hydrogen bonding distance. In this case, the problem can be fixed by flipping the imidazole ring by 180°. The flip will put the NE atom in the place of CD, allowing for the possibility of a hydrogen bond with the neighbouring oxygen. The fit of the histidine to the electron density map will remain good.

11.5 Prove

The idea behind Prove (Pontius *et al.*, 1996) is that poorly modeled regions of a structure can be identified by their irregular atomic volumes, having been squeezed too close or separated oo far from neighbouring atoms due to the constraints of ill fitting electron density. Prove calculates atomic volumes of buried atoms within the query structure and evaluates their deviations from standard values by the volume Z-score. A library of 64 well-refined, high resolution protein structures was used to derive the standard volumes of atoms. Atoms and their volumes are classified as belonging to one of 23 chemical types (e.g. methyl group *versus* methylene group). The algorithm is considered a means of cross-validation because atomic volumes are not directly restrained in refinement procedures.

In the 1ja3 example, we find a Z-score of 0.144 indicating an average atom size only slightly larger than standard (Figure 11.6, left panel, dot symbol). This value falls within the range expected for structures of this resolution (gray cone) and by itself, is not a cause for concern. However, the Z-score RMS of 1.85 is well outside the range expected for 3.0 Å resolution structures, signifying there are regions of the structure with very high atomic volumes and other regions of very low atomic volumes. This large RMS value should signal the crystallographer to examine Z-scores for individual residues. The plot, however, is relatively uninformative; there are no localized regions of high deviation that one could use as a focus for attention. Instead, deviations appear of equal magnitude throughout the structure. One might expect that the maximum Z-score deviations per residue would be significantly higher in the region 204–230, as this region has been repeatedly flagged as a trouble region by all the previously described structure validation tools. In this case, Prove appears to be useful as an indicator of overall structure quality, but is less helpful in pinpointing troubled areas. After rebuilding the structure as described in previous sections, these deviations decrease overall from an average of 1.85–1.37 (Figure 11.6 right panel, cross symbol).

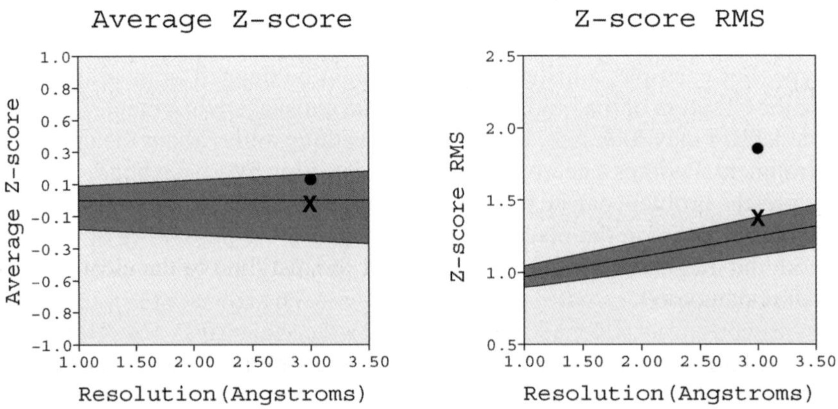

Fig. 11.6 Prove Z-Score diagrams for the model of 1ja3 as deposited with the PDB (dot symbol) and after revision (cross symbol).

12

General remarks

While writing this book, several things came to mind, which I found important enough to be mentioned, but which I could not quite fit into the context of the various chapters. Other things that are mentioned in one of the chapters are important enough to be repeated and elaborated upon in greater depth. In the following paragraphs I have tried to compile a short list of 'things I also wanted to say'.

12.1 How many refinement cycles do I need?

There is no general answer to this question. Too few cycles lead to incomplete convergence and four or six seems to be the minimum even in very well-behaved refinements. Too many cycles will not hurt the refinement but can waste computing time.[1] I usually use 10 cycles; less if a refinement converges very well, and more when needed. You can tell that a refinement converges when the values for 'Mean shift/esd' and 'Maximum' become very small[2] (ideally 0.000, but 0.01 or even 0.1 is acceptable in early stages of the refinement).

If, for some reason, you started a refinement with very many cycles—for example 100—and you see after 20 cycles that the refinement has converged already, there is no need to wait for the remaining 80 cycles to be finished. During every refinement cycle, SHELXL checks whether there is a file `name.fin`. When such a file is found, SHELXL deletes it and, after completing the current refinement cycle, instead of calculating further refinement cycles, it continues with the final structure-factor calculations, etc. and finishes the refinement regularly. This method is particularly useful for refinements of large structures like proteins. You can start a 100-cycle refinement before you go home in the evening. The next morning when you come back, you generate a .fin file and after a cup of coffee the refinement will be done.

12.2 What to do with NPD atoms?

In some of the examples in this book (e.g. 6.3.1 or 8.3.1), as well as most probably in your own practice as crystallographers, you have seen some atoms 'go NPD'. NPD stands for 'non-positive definite' and refers to a thermal ellipsoid with one or more of the three half-axes of the anisotropic displacement ellipsoid possessing a negative

[1] For small structures refined on a modern computer, one refinement cycle can take less than a second, so that computing time is hardly an issue nowadays.

[2] For CGLS refinement the value to be watched is called 'Max. shift'.

value. This is physically meaningless and a model containing an atom of this kind is not publishable. [3]

The answer to the question of how to treat NPD atoms depends largely on the reason why the atom is non-positive definite. Some people think that atoms that cannot be refined anisotropically should not be refined anisotropically, meaning that if an atom is listed as non-positive definite, it should be refined isotropically, as the data do not justify the anisotropic model. In some cases this is certainly true, especially when the crystal was of poor quality and gave rise to a noisy dataset of only very low resolution (e.g. 1.1 Å or worse). Usually in such a case, several or even most atoms are NPD and many others show pathologically shaped displacement ellipsoids. Sometimes, however, NPD atoms are observed with good high-resolution data and, frequently, something can be done to allow a full anisotropic model without NPD atoms.

If an atom is NPD because of incorrectly assigned element type (carbon instead of sulfur or so) or due to an unresolved disorder, it usually suffices to correct the error to rectify the situation. In other, more difficult situations, constraints and restraints can help.

Especially in twinned structures or structures with global pseudo-symmetry, but sometimes also in well-resolved disorders where a light atom of one component is located close to a heavier atom of the other component, strong correlation effects among the parameters of atoms related by the twin law, pseudo-symmetry or disorder cause the thermal ellipsoids of some atoms to 'give in' under the pressure of the other atoms, figuratively speaking. In such situations the use of similarity and rigid-bond restraints on anisotropic displacement parameters can do miracles.[4] Sometimes you may have to change the standard uncertainties to smaller values to make the restraints stronger (say `SIMU 0.01` and `DELU 0.005`) or combine `SIMU` and `DELU` with `ISOR`, which treats an atom as approximately isotropic (for a full description of these restraints see Section 2.6.2 and Figure 2.2).

If restraints do not help, it can be necessary to constrain two (or more) atoms to possess the same anisotropic displacement parameters, using the `EADP` constraint. This has been done in Examples 6.3.1 and 8.3.1 and, if there is a comprehensible reason for the atom to show NPD behaviour, can be quite adequate. If restraints or constraints were used to refine a structure, this should, however, always be mentioned in the publication.

12.3 How many restraints may I use in a structure?

In the opinion of many people, among them reviewers and editors of scientific journals, the number of restraints in a structure should not exceed the number of refined parameters. This seems to be a somewhat arbitrary choice, as restraints are

[3] At least it should not be published—sometimes it is surprising what seems to be 'publishable' to referees and editors.

[4] As mentioned in Chapter 5, this is one of the reasons why disorders should always be refined with the help of restraints on geometry (`SAME`, `SADI`) and displacement parameters (`SIMU`, `DELU`).

treated as additional data in the refinement (see Equation 2.6) and not as parameters. It would appear reasonable to demand the number of restraints to remain significantly lower than the number of independent reflections to make sure that you refine your structure mainly against your diffraction data and not against a large set of restraints, but I cannot see why a crystal structure refinement with more restraints than refined parameters cannot give rise to a sensible and publishable model.

Of course it always depends on what restraints are used and how they are used. As described in Chapter 2 there are two kinds of restraints: direct and relative ones. The former restrain a variable (e.g. an interatomic distance) to a certain target value, while the latter relate variables within the model without imposing any outside values. Relative restraints generally exert much milder influence on the model than direct ones, and even a very large number of justified `SADI` and `DELU` restraints cannot jeopardize the ethical integrity of the model. This may be entirely different with direct restraints, and the use of many strong `DFIX` and `ISOR` commands to make the model look the way the chemist wants it to, can indeed result in questionable crystal structures.[5]

In general, restraints must be applied with great care and only if justified. When appropriate, however, they should be employed without hesitation, and having more restraints than parameters in a refinement is nothing to be ashamed of.

12.4 Coordination geometries of some cations

Some cations have a characteristic coordination geometry that can help to identify them. Pt^{2+}, for example, is found almost exclusively fourfold coordinated with the four ligands lying in a common plane with the metal atom, while Pt^{4+} prefers octahedral geometry. On the other hand, other cations like lead or molybdenum are rather variable in their geometrical behaviour and it is not all that helpful to memorize all possible geometries. Therefore, the following table is incomplete and contains only a small selection of common cations with their most important coordination geometries. More commonly observed coordination numbers and geometries are printed in boldface, very rarely observed ones have not been included.

Ion	Coordination number	Coordination geometry
Li^+	4	tetrahedral
	6	**octahedral**
Na^+	4	tetrahedral
	6	**octahedral**
K^+	4	tetrahedral
	6	**octahedral**
	>6	various geometries

[5] George Sheldrick says 'with the right restraints, you can fit an elephant to any data'.

Ion	Coordination number	Coordination geometry
Mg^{2+}	4	tetrahedral
	6	**octahedral**
Ca^{2+}	**6**	**octahedral**
	>6	various geometries
Al^{3+}	**4**	**tetrahedral**
	5	trigonal bi-pyramidal
	6	**octahedral**
Ga^{3+}/In^{3+}	**4**	**tetrahedral**
	5	various geometries
	6	**octahedral**
Si^{4+}	3	planar
	4	**tetrahedral**
Ti^{+}/Ti^{2+}	**6**	**octahedral**
Ti^{3+}	3	planar
	5	trigonal bi-pyramidal
	6	**octahedral**
Ti^{4+}	**4**	**tetrahedral**
Zr^{4+}/Hf^{4+}	**6**	**octahedral**, trigonal prismatic
	>6	various geometries
V / Nb / Ta in all ox. states	**6**	**octahedral**
Cr^{3+}	**6**	**octahedral**
Cr^{4+}	**4**	**tetrahedral**
Mn^{+}/Mn^{4+}	**6**	**octahedral**
$Mn^{5+}/Mn^{6+}/Mn^{7+}$	**4**	**tetrahedral**
Co^{3+}	**6**	**octahedral**
Pd^{2+}/Pt^{2+}	**4**	**planar**
Pd^{4+}/Pt^{4+}	**6**	**octahedral**
Cu^{+}	**4**	**tetrahedral**
Cu^{2+}	**4**	**tetrahedral, planar**
	6	**octahedral** (Jahn-Teller)
Ag^{+}/Au^{+} (also Hg^{2+})	**2**	**linear**
$Ag^{2+}/Au^{2+}/Au^{3+}$	**4**	**planar**
Zn^{2+}	**4**	**tetrahedral**, planar
	6	octahedral
Cd^{2+}	4	tetrahedral
	6	**octahedral**

12.5 Some typical bond lengths

Below are several tables of some of the more common covalent bond distances, all given in Ångströms.

Single bonds

B	**C**	**N**	**O**	**F**	**Si**	**P**	**S**	**Cl**	**Br**	
1.62	1.58	1.49	1.37	1.32	1.98	1.94	1.81	1.74	1.89	**B**
	1.54	1.47	1.43	1.39	1.87	1.85	1.83	1.79	1.95	**C**
		1.45	1.41	1.36	1.74	1.70	1.69	1.75	2.14	**N**
			1.48	1.42	1.64	1.62	1.57	1.70	1.65	**O**
				1.42	1.56	1.57	1.54	1.64	1.76	**F**
					2.34	2.25	2.13	2.02	2.17	**Si**
						2.22	2.12	2.04	2.22	**P**
							2.07	2.01	2.24	**S**
								1.99	2.14	**Cl**
									2.29	**Br**

	C(sp^3)	**C(sp^2)**	**C(sp)**	**N(sp^3)**	**N(sp^2)**	**O**	**S**	**F**	**Cl**	**Br**
C(sp^3)	1.54	1.51	1.46	1.47	1.45	1.43	1.83	1.39	1.79	1.95
C(sp^2)		1.47	1.43	1.43	1.40	1.35	1.76	1.35	1.73	1.85
C(sp)			1.37	1.33	1.33	1.26		1.2	1.63	1.79

Double bonds

	C	**N**	**O**	**P**	**S**
C	1.34	1.29	1.21	1.67	1.63
N		1.25	1.22	1.55	1.52
O			1.21	1.47	1.43
P				2.03	1.92

	C(sp^2)	**C(sp)**	**N(sp^2)**	**O**	**S**
C(sp^2)	1.34	1.32	1.29	1.21	1.70
C(sp)		1.29	1.20	1.17	1.56

Triple bonds

	C	N	O	P	S
C	1.20	1.16	1.13	1.53	1.47
N		1.10	1.11		

12.6 Resolution tables

From Bragg's law, $\lambda = 2d \sin \Theta$, the relationship between the resolution d and the angle 2Θ can be computed very easily for any given wavelength λ. Below is a table that contains pairs of d [in Å] versus 2Θ [in °] for the two most commonly used anode materials, copper and molybdenum.

Mo Radiation: λ = 0.71073 Å

2Θ	***d***	***d***	**2Θ**
5	8.15	0.70	61.0
10	4.08	0.75	56.6
15	2.72	0.80	52.7
20	2.05	0.85	49.4
25	1.66	0.90	46.5
30	1.39	0.95	43.3
35	1.18	1.00	41.6
40	1.04	1.10	37.7
45	0.93	1.20	34.8
50	0.84	1.30	32.0
55	0.77	1.40	29.4
60	0.71	1.50	27.4
		2.00	20.5

Cu-Radiation: λ = 1.5418 Å

2Θ	***d***	***d***	**2Θ**
5	17.67	0.80	149.0
10	8.85	0.85	130.2
20	4.44	0.90	117.9
30	2.98	0.95	108.5
40	2.25	1.00	100.0
50	1.82	1.10	89.0
60	1.54	1.20	79.9

Cu-Radiation: λ = 1.5418 Å

2Θ	***d***	***d***	**2Θ**
70	1.34	1.30	72.7
80	1.20	1.40	66.8
90	1.09	1.50	61.9
100	1.01	2.00	45.3
110	0.94	3.00	29.8
120	0.89	4.00	22.2
130	0.85	5.00	17.7
140	0.82		
150	0.80		

References

Ackerhans, C., Böttcher, P., Müller, P., Roesky, H. W., Usón, I., Schmidt, H. G., and Noltemeyer, M. (2001). *Inorg. Chem.*, **50**, No. 15, 3766–3773.

Adamchuk, J., Schrock, R. R., Tonzetich, Z. J. and Müller, P. (2006). *Organometallics*, **25**, 2364–2373.

Allen, F. H. (2002). *Acta Crystallogr.*, **B58**, 380–388.

Andersson, K. M. and Hovmöller, S. (1998). *Z. Kristallogr.*, **213**, 369–373.

Arnberg, L., Hovmöller, S., and Westerman, S. (1979). *Acta Crystallogr.*, **A35**, 497–499.

Bader, R. F. W. (1990). *Atoms in Molecules, a Quantum Theory*, Oxford: Clarendon Press.

Bai, G., Müller, P., Roesky, H. W., and Usón, I. (2000). *Organometallics*, **19**, 4675–4677.

Berisio, R., Lamzin, V. S., Sica, F., Wilson, K. S., Zagari, A., and Mazzarella, L. (1999) . *J. Mol. Biol.*, **292**, 845–854.

Bernardinelli, G. and Flack, H. D. (1985). *Acta Crystallogr.*, **A41**, 500–511.

Boese, R. (1999). Der Vergleich von Röntgenstrukturdaten—Fehler und Artefakte *In:* Workshop Struktrubestimmung—Datenbanken—Molecular Modelling, November 27–30 Frankfurt/Main (Abstract book of the 1999 KSAM-Meeting in Frankfurt/Main, Germany).

Breyer, W. A., Kingston, R. L., Anderson, B. F., and Baker, E. N. (1999). *Acta Crystallogr.*, *D***55**, 129–138.

Bruker (1999). GEMINI. Bruker AXS Inc., Madison, WI.

Bruker (2001). SAINT. Bruker AXS Inc., Madison, WI.

Brünger, A. T. (1992). *Nature*, **355**, 472–474.

Celli, A. M., Donati, D., Fonticelli, F., Roberts-Blaming, S. J., Kalaji, M., and Murphy, P.J. (2001). *Org. Lett.*, **3**, 3573–3574.

Clegg, W. (1982). *Acta Crystallogr.*, **B38**, 1648–1649.

Cochran, W. and Lipson, H. (1966). *In:* W. L. Bragg, ed. *The Determination of Crystal Structures*. Ithaca, NY: Cornell University Press, pp. 323–330.

Collaborative Computational Project Number 4 (1994). *Acta Crystallogr.*, **D50**, 760–763.

Colovos, C. and Yeates, T. O. (1993). *Protein Sci.*, **2**, 1511–1519.

Cooper, R. I., Gould, R. O., Parsons, S., and Watkin, D. J. (2002). *J. Appl. Crystallogr.*, **35**, 168–174.

Cruickshank, D. W. (1970). *In:* F. R. Ahmed, S. R. Hall, C. P. Huber, eds. *Crystallographic Computing*. Copenhagen: Munksgaard Publ. pp. 187–197.

Cruickshank, D. W. J. and McDonald, W. S. (1967). *Acta Crystallogr.*, **23**, 91–11.

Dauter, Z. (2003). *Acta Crystallogr.*, **D59**, 2004–2016.

DeLaMatter, D., McCullough, J. J., and Calvo, C. (1973). *J. Phys. Chem.*, **77**, 1146–1148.

DeLano (2002). *The PyMOL Molecular Graphics System.* DeLano Scientific, San Carlos, CA, USA.

Didisheim, J. J. and Schwarzenbach, D. (1987). *Acta Crystallogr.*, **A43**, 226–232.

Dimasi, N., Sawicki, M. W., Reineck, L. A., Li, Y., Natarajan, K., Marguiles, D. H., and Mariuzza, R.A. (2002). *J. Mol. Biol.*, **320**, 573–585.

Duisenberg, A. J. M. (1992). *J. Appl. Crystallogr.*, **25**, 92–96.

Duisenberg, A. J. M., Kroon-Batenburg, L. M. J., and Schreurs, A. M. M. (2003). *J. Appl. Crystallogr.*, **36**, 220–229.

Dunitz, J. D., Maverick, E. F., and Trueblood, K. N. (1988). *Angew. Chem. Int. Ed. Engl.*, **27**, 880–895.

Einsle, O. F., Tezcan, A., Andrade, S. L. A., Schmid, B., Yoshida, M., Howerd, J. B., and Rees, D. C. (2002). *Science*, **297**, 1696–1700.

Emsley, P. and Cowtan, K. (2004). *Acta Crystallogr.*, **D60**, 2126–2132.

Engh, R. and Huber, R. (1991). *Acta Crystallogr.*, **A47**, 392–400.

Esnouf, R. M. (1999). *Acta Crystallogr.*, **D55**, 938–940.

Esposito, L., Vitagliano, L., and Mazzarella, L. (2002). *Protein Pept. Lett.*, **9**, 95–105.

Farrugia, L. J. (2000). IUCRVAL, University of Glasgow, Scotland.

Flack, H. D. (1983). *Acta Crystallogr.*, **A39**, 876–881.

Flack, H. D. and Schwarzenbach, D. (1988). *Acta Crystallogr.*, **A44**, 499–506.

Friedel, G. (1928). *Leçons de Cristallographie*. Paris: Berger-Levrault.

Fukuyo, M., Hirotsu, K., and Higuchi, T. (1982). *Acta Crystallogr.*, **B38**, 640–643.

Genick, U. K., Soltis, S. M., Kuhn, P., Canestrelli, I. L., and Getzoff, E. D. (1998). *Nature*, **392**, 206–209.

Gerke, R., Fitjer, L., Müller, P., Usón, I., Kowski, K., and Rademacher, P. (1999). *Tetrahedron*, **55**, 14429–14434.

Giacovazzo, C., ed. (2002). *Fundamentals in Crystallography*. 2nd edn. Oxford: Oxford University Press.

Grosse-Kunstleve, R. W., and Adams, P. D. (2001). *J. Appl. Crystallogr.*, **35**, 477–480.

Hall, S. R., Allen, F. H., and Brown, I. D. (1991). *Acta Crystallogr.*, **A47**, 655–685.

Harlow, R. L. (1996). *J. Res. Natl. Inst. Stand. Technol.*, **101**, 327–339.

Hatop, H., Schiefer, M., Roesky, H. W., Herbst-Irmer, R., and Labahn, T. (2001). *Organometallics*, **20**, 2643–2646.

Heine, A., DeSantis, G., Luz, J. G., Mitchell, M., Wong, C. H., and Wilson, I. A. (2001). *Science*, **294**, 369–374.

Herbst-Irmer, R. and Sheldrick, G. M. (1998). *Acta Crystallogr.*, **B54**, 443–449.

Herbst-Irmer, R. and Sheldrick, G. M. (2002). *Acta Crystallogr.*, **B58**, 477–481.

Hirshfeld, F. L. (1976). *Acta Crystallogr.*, **A32**, 239–244.

Hirschfeld, F. L. and Rabinovich, D. (1973). *Acta Crystallogr.*, **A29**, 510–513.

Hoenle, W. and von Schnering, H. G. (1988). *Z. Krist.*, **184**, 301–305.

Hooft, R. W. W., Vriend, G., Sander, C., and Abola, E. E. (1996). *Nature*, **381**, 272–277.

Howard, E. I., Sanishvili, R., Cachau, R. E., Mitschler, A., Chevrier, B., Barth, P., Lamour, V., Van Zandt, M., Sibley, E., Bon, C., Moras, D., Schneider, T. R., Joachimiak, A., and Podjarny, A. (2004). *Proteins*, **55**, 792–804.

Hutmacher, H. M., Fritz, H. G., and Mussow, H. (1975). *Angew. Chem.*, **14**, 180–181.

Jameson, G. B. (1982). *Acta Crystallogr.*, **A38**, 817–820.

Jancarick, J. and Kim, S.-H. (1991). *J. Appl. Crystallogr.*, **24**, 409–411.

Johnson, C. K. (1976). ORTEP-II, Oak Ridge National Laboratory, Tennessee, USA.

Jones, A. T., Zou, J. Y., Cowan, S.W., and Kjelgaard, M. (1991). *Acta. Crystallogr.*, **A47**, 110–119.

Jones, P. G., Vancea, F., and Herbst-Irmer, R. (2002). *Acta Crystallogr.*, **C58**, o665–o668.

Kahlenberg, V. (1999). *Acta Crystallogr.*, **B55**, 745–751.

Kahlenberg, V. and Messner, T. (2001). *J. Appl. Crystallogr.*, **34**, 405–405.

Kapon, M. and Herbstein, F. H. (1995). *Acta Crystallogr.*, **B51**, 108–113.

Kidera, A., Inaka, K., Matsushima, M., and Go, N. (1992). *J. Mol. Biol.*, **225**, 477–486.

König, V., Vértesy, L., and Schneider, T. R. (2003). *Acta Crystallogr.*, **D59**, 1737–1743.

Larson, A. C. (1970). *In*: F. R. Ahmed, S. R. Hall, and C. P. Huber, eds. *Crystallographic Computing*. Copenhagen: Munksgaard Publ., pp. 291–294.

Laskowski, R. A., MacArthur, M. W., Moss, D. S., and Thornton, J. M. (1993). *J. Appl. Crystallogr.*, **26**, 283–291.

Le Page, Y. (1982). *J. Appl. Crystallogr.*, **15**, 255–259.

Le Page, Y. (1987). *J. Appl. Crystallogr.*, **20**, 264–269.
Le Page, Y. (1988). *J. Appl. Crystallogr.*, **21**, 983–984.
Li, J., Burgett, A. W. G., Esser, L., Amezcua, C., and Harran, P. G. (2001). *Angew. Chem. Int. Ed.*, **40**, 4770–4773.
Lüthy, R., Bowie, J. U., and Eisenberg, D. (1992). *Nature* **356**, 83–85.
MacArthur, M. W., and Thornton, J. M. (1996). *J. Mol. Biol.*, **264**, 1180–1195.
Marsh, R. E. and Spek, A. L. (2001). *Acta Crystallogr.*, **B57**, 800–805.
McRee, D. E. (1999). *J. Struct. Biol.*, **125**, 156–165.
Merrit, E. A. (1999). *Acta Crystallogr.*, **D55**, 1109–1117.
Moews, P. C. and Kretsinger, R. H. (1975). *J. Mol. Biol.*, **91**, 201–228, 1975.
Morris, J. M. and Bricogne, G. (2003). *Acta Crystallogr.*, **D59**, 615–617.
Müller, P. (2001). *Probleme der modernen hochauflösenden Einkristall-Röntgenstrukturanalyse*, Thesis (PhD), University of Göttingen, Germany.
Müller, P. (2005). MoO is no Schmu. *In:* Annual meeting of the American Crystallographic Association, May 28–June 2, Orlando, FL.
Müller, P., Sawaya, M. R., Pashkov, I., Chan, S., Nguyen, C., Wu, Y., Perry, L. J., and Eisenberg, D. (2005). *Acta Crystallogr.*, **D61**, 309–315.
Müller, P., Usón, I., Hensel, V., Schlüter, A. D., and Sheldrick, G. M. (2001). *Helv. Chim. Acta*, **84**, 778–785.
Müller, P., Usón, I., Prust, J., and Roesky, H. W. (2000). *Acta Crystallogr.*, **C56**, 1300.
Otwinowski, Z., and Minor, W. (1997). *In:* R. M. Sweet and C. W. Carter Jr., eds. *Methods in Enzymology*, volume 276, Orlando, FL: Academic Press, pp. 307–326.
Padilla, J. E. and Yeates, T. O. (2003). *Acta Crystallogr.*, **D59**, 1124–1130.
Parkin, G. (1993). *Chem. Rev.*, **93**, 887–911.
Pauling, L. (1947). *J. Am. Chem. Soc.*, **69**, 542–553.
Perrakis, A., Morris, R. M., and Lamzin, V. S. (1999). *Nat. Struct. Biol.*, **6**, 458–463.
Peterson, S. W., Gebert, E., Reis, A. H., Druyan, Jr. M. E., Mason, G. W., and Peppard, D. F. (1977). *J. Phys. Chem.*, **81**, 466–471.
Pflugrath, J. W., Wiegand, G., Huber, R., and Vertesy, L. (1986). *J. Mol. Biol.*, **189**, 383–386.
Pontius, J., Richelle, J., and Wodak, S. J. (1996). *J. Mol. Biol.*, **264**, 121–136.
Pratt, C. S., Coyle, B. A., and Ibers, J. A. (1971). *J. Chem. Soc.*, **A**, 2146–2151.
Press, W. H., Teukolsky, S. A., Vetterling, W. T., and Flannery, B. P. (2002). *Numerical Recipes in C++*, Cambridge: Cambridge University Press.
Ramakrishnan, C. and Ramachandran, G. N. (1965). *Biophys. J.*, **5**, 909–993.
Rees, D. C. (1980). *Acta Crystallogr.*, **A36**, 578–581.
Rennekamp, C., Müller, P., Prust, J., Wessel, H., Roesky, H. W., and Usón, I. (2000). *Eur. J. Inorg. Chem.*, **5**, 1861–1868.
Rollett, J. S. (1970). *In:* F. R. Ahmed, S. R. Hall, and C. P. Huber, eds. *Crystallographic Computing*. Copenhagen: Munksgaard Publ., pp. 167–181.
Rosenfield, R. E., Trueblood, K. N., and Dunitz, J. D. (1978). *Acta Crystallogr.*, **A34**, 828–829.
Rudolph, M. G., Kelker, M. S., Schneider, T. R., Yeates, T. O., Oseroff, V., Heidary, D. K., Jennings, P. A., and Wilson, I. A. (2003). *Acta Crystallogr., D***59**, 290–298.
Saenger, W., Betzel, C., Hingerty, B., and Brown, G. M. (1982). *Nature*, **296**, 581–583.
Schmidt, A., and Lamzin, V. S. (2002).*Curr. Opin. Struct. Biol.*, **12**, 698–703.
Schneider, T. R. (1996a). *In:* E. Dodson, M. Moore, S. Bailey, eds. *Proceedings of the CCP4 Study Weekend*, Warrington: Daresbury Laboratories, pp. 133–144.
Schneider, T. R. (1996b). *Röntgenkristallographische Untersuchung der Struktur und Dynamik einer Serinprotease*, Thesis (PhD), Universität München, Germany.
Schomaker, V. and Trueblood, K. N. (1968). *Acta Crystallogr.*, **B24**, 63–76.

Schröder Leiros, H.-K., McSweeney, S. M., and Smalås, A. O. (2001). *Acta Crystallogr.*, **D58**, 1307–1313.

Sevcík, J., Lamzin, V. S., Dauter, Z., and Wilson, K. S. (2002). *Acta Crystallogr.*, **D57**, 488–597.

Sheldrick, G. M. (1990). *Acta Crystallogr.*, **A46**, 467–473.

Sheldrick, G. M. (1992). *In:* D. Moras, A. D. Podjarny, J. C. Thierry, eds. *Crystallographic Computing*, volume 5. Oxford: Oxford University Press, pp. 145–157.

Sheldrick, G. M. (1997a). SADABS, University of Göttingen.

Sheldrick, G. M. (1997b). SHELXL-97, University of Göttingen.

Sheldrick, G. M. (2001). XPREP 6.12, SHELXTL, Bruker-AXS.

Sheldrick, G. M. (2002). TWINABS, University of Göttingen.

Sheldrick, G. M. (2003). CELL_NOW, University of Göttingen.

Sheldrick, G. M. and Schneider, T. R. (1997). *In:* R. M. Sweet and C. W. Carter Jr., eds. *Methods in Enzymology*, volume 277, Orlando, FL: Academic Press, pp. 319–343.

Sluis, P. van der and Spek, A. L. (1990). *Acta Crystallogr.*, **A46**, 194–201.

Sparks, R. A. (1997). Twinning—programs for indexing, structure refinement and determining the relationship between the twin components. *In: Annual Meeting of the American Crystallographic Association*, July 19–25, St. Louis, MI.

Spek, A. L. (2003). *J. Appl. Crystallogr.*, **36**, 7–13.

Spek, A. L. (2006). *PLATON: A Multipurpose Crystallographic Tool*. Utrecht University.

Trueblood, K. N., Bürgi, H.-B., Burzlaff, H., Dunitz, J. D., Gramaccioli, C. M., Schulz, H. H., Shmueli, U., and Abrahams, S. C. (1996). *Acta Crystallogr.*, **A52**, 770–781.

Trueblood, K. N. and Dunitz, J. D. (1983). *Acta Crystallogr.*, **B39**, 120–133.

Usón, I., Pohl, E., Schneider, T. R., Dauter, Z., Schmidt, A., Fritz, H. J., and Sheldrick, G. M. (1999). *Acta Crystallogr.*, **D55**, 1158–1167.

Usón, I. and Sheldrick, G. M. (1999). *Curr. Opin. Struct. Biol.*, **9**, 643–648.

Vrielink, A. and Sampson, N. (2003). *Curr. Opin. Struct. Biol.*, **13**, 109–715.

Watkin, D. (1994). *Acta Crystallogr.*, **A50**, 411–437.

Word, J. M., Lovell, S. C., LaBean, T. H., Tayler, H. C., Zalis, M. E., Presley, B. K., Richardson, J. S., and Richardson, D. C. (1999). *J. Mol. Biol.*, **285**, 1711–1733.

Yeates, T. (1997). *In:* R. M. Sweet and C. W. Carter Jr., eds. *Methods in Enzymology*, volume 276, Orlando, FL: Academic Press, pp. 344–358.

Yu, P., Müller, P., Said, M. A., Roesky, H. W., Usón, I., Bai, G., and Noltemeyer, M. (1999). *Organometallics*, **18**, 1669–1674.

Websites (All accessed November 24 2005)

ARP/wARP: www.embl-hamburg.de/ARP/

BobScript: www.strubi.ox.ac.uk/bobscript/

Coot. www.ysbl.york.ac.uk/~emsley/coot/

ccp4. www.ccp4.ac.uk/

ERRAT: www.nihserver.mbi.ucla.edu/ERRATv2/

IUCr validation criteria: www.journals.iucr.org/services/cif/checking/autolist.html

IUCr checkCIF: www.journals.iucr.org/services/cif/checking/checkform.html

Numerical Recipes: www.nr.com

Ortep: www.ornl.gov/sci/ortep/

Ortep for Windows: www.chem.gla.ac.uk/~louis/software/ortep3/

Parvati: www.bmsc.washington.edu/parvati/parvati.html

PDB Validation Server: www.deposit.pdb.org/validate/
PLATON: www.xraysoft.chem.uu.nl
PROCHECK: www.biochem.ucl.ac.uk/~roman/procheck/procheck.html
PROVE: www.biotech.ebi.ac.uk:8400/doc/prove/prove.html
PyMOL: www.pymol.sourceforge.net/
SAVS: www.nihserver.mbi.ucla.edu/SAVS/
SHELX: www.shelx.uni-ac.gwdg.de/SHELX
Twin Server: www.doe-mbi.ucla.edu/Services/Twinning
Verify3D: www.nihserver.mbi.ucla.edu/Verify_3D/
WHAT_CHECK: www.swift.cmbi.ru.nl/gv/whatcheck/
WinGX: www.chem.gla.ac.uk/~louis/software/wingx/
Xfit: www.sdsc.edu/CCMS/Packages/XTALVIEW/xtalview.html

Further Reading

This book is meant to help the already somewhat advanced crystallographer with common refinement problems and not to give an exhaustive overview over crystal structure refinement or a general introduction into the field of crystallography. There is a variety of excellent textbooks and articles available, and the following reading provides deeper insight. The inclined reader may turn to the references below in order to gain a more sound knowledge if he or she wishes to do so.

W. Massa (2004). *Crystal Structure Determination.* 2nd edn. New York: Springer.

This is the ideal book for the beginner. In the excellent translation by R. O. Gould, the Massa explains all the basics from symmetry in real and reciprocal space, over generation of X-rays and other practical aspects, to structure solution and refinement. Many experienced teachers of introductory crystallography classes recommend this book to their students, and it was the German original version of this book, that helped me to understand crystallography when I was a beginner myself.

W. Clegg (1998). *Crystal Structure Determination*, Oxford: Oxford University Press.

If you want it in a nutshell then this book is for you. On less than 100 pages Clegg covers the most important basic aspects of X-ray structure determination.

C. Giacovazzo, ed. (2002). *Fundamentals of Crystallography.* 2nd edn. Oxford: Oxford University Press.

This is the ideal book to deepen the knowledge one has gained from a beginner's book like the Massa. Most important aspects of crystallography are explained in an understandable fashion and every serious crystallographer should read this book.

W. Clegg, ed. (2001). *Crystal Structure Analysis*, Oxford: Oxford University Press.

The content of this book is based on material from an 'Intensive Course in X-Ray Structure Analysis'. It is a practical approach to crystallography and describes crystal growth (not many books do), several data collection techniques, methods of structure solution and refinement as well as the interpretation of crystallographic results and crystallographic data bases.

J. P. Glusker and K. N. Trueblood (1985). *Crystal Structure Analysis—A Primer*. 2nd edn. Oxford: Oxford University Press.

This book is a classic. Intended mainly for biologists, it explains everything from the X-ray diffraction pattern to the three dimensional structure. It uses clear and understandable language, not too many formulae and has excellent illustrations.

G. H. Stout and L. H. Jensen (1989). *X-Ray Structure Determination*. 2nd edn. New York: Wiley-Interscience.

With many good graphics, this book explains it all. From the beginning (generation and diffraction of X-rays) to the end (errors and ambiguities of the method), almost everything you ever wanted to know (and more) about the method of X-ray structure determination is explained here. Unfortunately the book is a little outdated and more recent developments like area detectors are not mentioned.

H. Lipson & W. Cochran (1966). *The Determination of Crystal Structures.* 3rd edn. Ithaca, NY: Cornell University Press.

This is the third volume in the series *The Crystalline State* edited by Sir Lawrence Bragg. Like volumes one and two (volume 1: L. Bragg (3rd ed. 1966), *The Crystalline State.* Volume 2: R. W. James (1965), *The Optical Principles of the Diffraction of X-Rays.*) this is still an amazing book. While it is both seriously outdated and out of print, many details of the method of X-ray structure determination that everyone takes for granted nowadays are presented as new ideas and thoroughly explained in a thrilling way. While I would definitely not recommend this book to a beginner, it can be a great pleasure and enjoyment for the expert to read in this book.

The following web sites are worth visiting [All accessed November 20 2005]

- *Kevin Cowtan's Tutorials* on Fourier transformation and on structure facture calculation are famous for their clarity and instructiveness. With his 'Fourier Duck' and 'Fourier Cat' in the Book of Fourier Cowtan has created a legend: www.ysbl.york.ac.uk/~cowtan/
- *Mike Sawaya's Tutorials* on various aspects of practical protein crystallography are uniquely understandable and many important programs are explained clearly and with many well chosen examples: www.doe-mbi.ucla.edu/~sawaya/tutorials/tutorials.html
- *Bernhard Rupp's Crystallography 101* is a more general introduction into crystallography. It is a very nice online textbook starting from the beginning: ruppweb.dyndns.org/Xray/101index.html
- *Eftichia Alexopoulos' and Fabio Dall' Antonia's SHELXTL Tutorial* is the ideal training for the beginner and is almost like a zeroeth chapter to this book. The basics of the programs XPREP, SHELXS, SHELXL and XP are explained from scratch and in detail: shelx.uni-ac.gwdg.de/tutorial/english/index.html

Index